Trigonometry Workbook For Dummies

by Mary Jane Sterling

Other *For Dummies* math titles:

Algebra For Dummies 0-7645-5325-9
Algebra Workbook For Dummies 0-7645-8467-7
Calculus For Dummies 0-7645-2498-4
Calculus Workbook For Dummies 0-7645-8782-x
Geometry For Dummies 0-7645-5324-0
Statistics For Dummies 0-7645-5423-9
Statistics Workbook For Dummies 0-7645-8466-9
TI-89 Graphing Calculator For Dummies 0-7645-8912-1 (also available for TI-83 and TI-84 models)
Trigonometry For Dummies 0-7645-6903-1

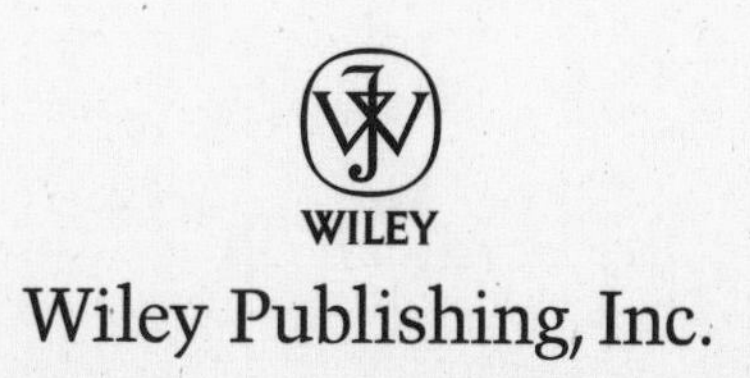

Wiley Publishing, Inc.

Trigonometry Workbook For Dummies®

Published by
John Wiley & Sons, Inc.
111 River St.
Hoboken, NJ 07030-5774
www.wiley.com

For general information on our other products and services, please contact our Customer Care Department within the U.S. at 877-762-2974, outside the U.S. at 317-572-3993, or fax 317-572-4002.

For technical support, please visit www.wiley.com/techsupport.

Wiley publishes in a variety of print and electronic formats and by print-on-demand. Some material included with standard print versions of this book may not be included in e-books or in print-on-demand. If this book refers to media such as a CD or DVD that is not included in the version you purchased, you may download this material at http://booksupport.wiley.com. For more information about Wiley products, visit www.wiley.com.

ISBN 978-0-7645-8781-8 (pbk); ISBN 978-0-470-22214-0 (ebk); ISBN 978-0-471-76403-8 (ebk)

Manufactured in the United States of America

10 9 8 7 6

1B/SQ/QW/QV/IN

About the Author

Mary Jane Sterling is also the author of *Algebra For Dummies, Trigonometry For Dummies, Algebra Workbook For Dummies, Algebra I CliffStudySolver*, and *Algebra II CliffStudySolver* (all published by Wiley). She has taught at Bradley University in Peoria, Illinois, for over 25 years.

Dedication

I would like to dedicate this book to my husband, Ted, for his good-natured patience and understanding during the tense times of this and other writing projects. I also dedicate this book to my three children — Jon, Jim, and Jane — who seem to get a kick out of having a mother who writes books about mathematics.

Author's Acknowledgments

I would like to thank Elizabeth Kuball for all her hard work on whipping this into shape; she has been great to work with. Thank you to David Herzog for his technical input. Also, thanks to Kathy Cox for seeing that I got another great project.

Publisher's Acknowledgments

We're proud of this book; please send us your comments through our Dummies online registration form located at `www.dummies.com/register/`.

Some of the people who helped bring this book to market include the following:

Acquisitions, Editorial, and Media Development

Project Editor: Elizabeth Kuball

Acquisitions Editor: Kathy Cox

Technical Editor: David Herzog

Editorial Manager: Michelle Hacker

Editorial Assistants: Hanna Scott, Melissa Bennett

Cover Photos: © Getty Images/Photodisc Blue

Cartoons: Rich Tennant (`www.the5thwave.com`)

Composition Services

Project Coordinator: Shannon Schiller

Layout and Graphics: Jonelle Burns, Andrea Dahl, Carrie A. Foster, Lauren Goddard, Denny Hager, Heather Ryan, Rashell Smith

Proofreaders: Vicki Broyles, Leeann Harney, Jessica Kramer, Carl Pierce, Dwight Ramsey

Indexer: Lynnzee Elze

Publishing and Editorial for Consumer Dummies

Kathleen Nebenhaus, Vice President and Executive Publisher

Kristin Ferguson-Wagstaffe, Product Development Director

Ensley Eikenburg, Associate Publisher, Travel

Kelly Regan, Editorial Director, Travel

Publishing for Technology Dummies

Andy Cummings, Vice President and Publisher

Composition Services

Debbie Stailey, Director of Composition Services

Contents at a Glance

Table of Contents

Introduction

What in the world is *trigonometry?* Well, for starters, trigonometry is *in* the world, *on* the world, and *above* the world — at least its uses are. Trigonometry started out as a practical way of finding out how far things are from one another when you can't measure them. Ancient mathematicians came up with a measure called an *angle,* and the rest is history.

So, what's my angle in this endeavor? (Pardon the pun.) I wanted to write this book because trigonometry just hasn't gotten enough attention lately. You can't do much navigating without trigonometry. You can't build bridges or skyscrapers without trigonometry. Why has it been neglected as of late? It hasn't been ignored as much as it just hasn't been the center of attention. And that's a shame.

Trigonometry is about angles, sure. You can't do anything without knowing what the different angle measures do to the different trig functions. But trigonometry is also about relationships — just like some of these new reality television shows. Did I get your attention? These relationships are nearly as exciting as those on TV where they decide who gets to stay and who gets to leave. The sine gets to stay and the cosecant has to leave when you know the identities and rules and apply them correctly. Trigonometry allows you to do some pretty neat things with equations and mathematical statements. It's got the power.

Another neat thing about trigonometry is the way it uses algebra. In fact, algebra is a huge part of trigonometry. Thinking back to my school days, I think I learned more about the finesse of algebra when doing those trig identities than I did in my algebra classes. It all fits together so nicely.

Whatever your plans are for trigonometry, you'll find the rules, the hints, the practice, and the support in this book. Have at it.

About This Book

This book is intended to cement your understanding — to give you the confidence that you do, indeed, know about a particular aspect of trigonometry. In each section, you'll find brief explanations of the concept. If that isn't enough, refer to your copy of *Trigonometry For Dummies,* your textbook, or some other trig resource. With the examples I give, you'll probably be ready to try out the problems for yourself and move on from there. The exercises are carefully selected to incorporate the different possibilities that come with each topic — the effect of different kinds of angles or factoring or trig functions.

Conventions Used in This Book

Reading any book involving mathematics can have an added challenge if you aren't familiar with the conventions being used. The following conventions are used throughout the text to make things consistent and easy to understand:

- New terms appear in *italic* and are closely followed by an easy-to-understand definition.
- **Bold** is used to highlight the action parts of numbered steps. **Bold** is also used on the answers to the example and practice problems to make them easily identifiable.
- Numbers are either written out as words or given in their numerical form — whichever seems to fit at the moment and cause the least amount of confusion.
- The *variables* (things that stand for some number or numbers — usually unknown at first) are usually represented by letters at the end of the alphabet such as x, y, and z. The *constants* (numbers that never change) are usually represented by letters at the beginning of the alphabet such as a, b, or c, and also by two big favorites, k or π. In any case, the variables and constants are *italicized* for your benefit.
- Angle measures are indicated with the word *degrees* or the symbol for a degree, °, or the word *radians*. The radian measures are usually given as numbers or multiples of π. If the angle measure is unknown, I use the variable x or, sometimes, the Greek letter Θ.
- I use the traditional symbols for the mathematical operations: addition, + ; subtraction, – ; multiplication, × or sometimes just a *dot* between values; and division, ÷ or sometimes a slash, / .

Foolish Assumptions

We all make foolish assumptions at times, and here are mine concerning you:

- You have a basic knowledge of algebra and can solve simple linear and quadratic equations. If this isn't true, you may want to brush up a bit with *Algebra For Dummies* or a textbook.
- You aren't afraid of fractions. FOF (fear of fractions) is a debilitating but completely curable malady. You just need to understand how they work — and don't work — and not let them throw you.
- You have a scientific calculator (one that does powers and roots) available so you can approximate the values of radical expressions and do computations that are too big or small for paper and pencil.
- You want to improve your skills in trigonometry, practice up on those topics that you're a little rusty at, or impress your son/daughter/boyfriend/girlfriend/boss/ soul mate with your knowledge and skill in trigonometry.

How This Book Is Organized

This book is organized into parts. Trigonometry divides up nicely into these groupings or parts with similar topics falling together. You can identify the part that you want to go to and cover as much if not all of the section before moving on.

Part I: Trying Out Trig: Starting at the Beginning

The study of trigonometry starts with angles and their measures. This is what makes trigonometry so different from other mathematical topics — you get to see what angle measures can do. For starters, I describe, pull apart, and inspect angles in triangles and circles. You get intimate with the circle and some of its features; think of it as becoming well-rounded. (Sorry, I just couldn't resist.)

One of the best things about trigonometry is how visual its topics are. You get to look at pictures of angles, triangles, circles, and sketches depicting practical applications. One of the visuals is the coordinate plane. You plot points, compute distances and slopes, determine midpoints, and write equations that represent circles. This is preparation for determining the values of the trig functions in terms of angles that are all over the place — angles that have positive or negative, very small or very large, degree or radian measures.

And, saving the best for last, I cover the right triangles. These triangles start you out in terms of the trig functions and are very user-friendly when doing practical applications.

Part II: Trigonometric Functions

The trig functions are unique. These six basic functions take a simple little angle measure, chew on it a bit, and spit out a number. How do they do that? That's what you find out in the chapters in this part. Each function has its own particular definition and inner workings. Each function has special things about it in terms of what angles it can accept and what numerical values it produces. You start with the right triangle to formulate these functions, and then you branch out into all the angles that can be formed going 'round and 'round the circle.

Part III: Trigonometric Identities and Equations

The trigonometric identities are those special equivalences that the six trig functions are involved with. These identities allow you to change from one function to another for your convenience, or just because you want to. You'll find out what the identities are and what to do with them. Sometimes they help make a complex expression much simpler. Sometimes they make an equation more manageable — and solvable. (Believe it or not, some people actually like to solve trig identities just for the pure pleasure of conquering the algebraic and trigonometric challenge they afford.)

In this part, I introduce you to the inverse trig functions. They undo what the original trig function did. These inverse functions are very helpful when solving trig equations — equations that use algebra to find out which angles make the statement true.

And, last but not least, you'll find the Law of Sines and Law of Cosines in this part. These two laws or equations describe some relationships between the angles and sides of a triangle — and then use these properties to find a missing angle measure or missing side of the triangle. They're most handy when you can't quite fit a right triangle into the situation.

Part IV: Graphing the Trigonometric Functions

The trig functions are all recognizable by their graphs — or, they will be by the time you finish with this part. The characteristics of the functions — in terms of what angle measures they accept and what values they spew out — are depicted graphically. Pictures are very helpful when you're trying to convince someone else or yourself what's going on.

The graphs of the trig functions are transformed in all the ways possible — shoved around the coordinate system, stretched out, squashed, and flipped. I describe all these possibilities with symbols and algebra and with the actual graph. Even if you're graphing functions with a graphing calculator, you really need to know what's going on so you can either decipher what's on your calculator screen or tell if what you have is right or wrong.

Part V: The Part of Tens

This is one of my favorite parts of this book. Here I was able to introduce some information that just didn't fit in the other parts — stuff I wanted to show you and couldn't have otherwise. You'll find some identities that fit special situations and all have a connection with the minus sign. You'll find everything you've always wanted to know about a circle but were afraid to ask. And, finally, I explore and lay bare for all to see the relationships between the angles and sides of a triangle.

Icons Used in This Book

To make this book easier to read and simpler to use, I include some icons that can help you find and fathom key ideas and information.

You'll find one or more examples with each section in this book. These are designed to cover the techniques and properties of the topic at hand. They get you started on doing the practice problems that follow. The solutions at the end of each chapter provide even more detail on how to solve those problems.

This icon appears when I'm thinking, "Oh, it would help if I could mention that. . . ." These situations occur when there's a particularly confusing or special or complicated step in a problem. I use this icon when I want to point out something to save you time and frustration.

Sometimes, when you're in the thick of things, recalling a particular rule or process that can ease your way is difficult. I use this icon when I'm mentioning something you'll want to try to remember, or when I'm reminding you of something I've covered already.

Do you remember the old *Star Trek* series in which the computer would say, "Warning, warning!" and alert Commander Kirk and the others? Think of this icon as being an alert to watch out for Klingons or any other nasty, tricky, or troublesome situation.

Where to Go from Here

Where do you start? You can start anywhere you want. As with all *For Dummies* books, the design is with you in mind. You won't spoil the ending by doing those exercises, first. You can open to a random page or, more likely, look in the table of contents or index for that topic that's been bugging you. You don't have to start at the beginning and slog your way through. All through the book, I reference preceding and later chapters that either offer more explanation or a place for further discovery.

There's a great companion book to this workbook called, just by coincidence, *Trigonometry For Dummies.* It has more detail on the topics in this workbook, if you want to delve further into a topic or get something clarified.

Part I

Trying Out Trig: Starting at the Beginning

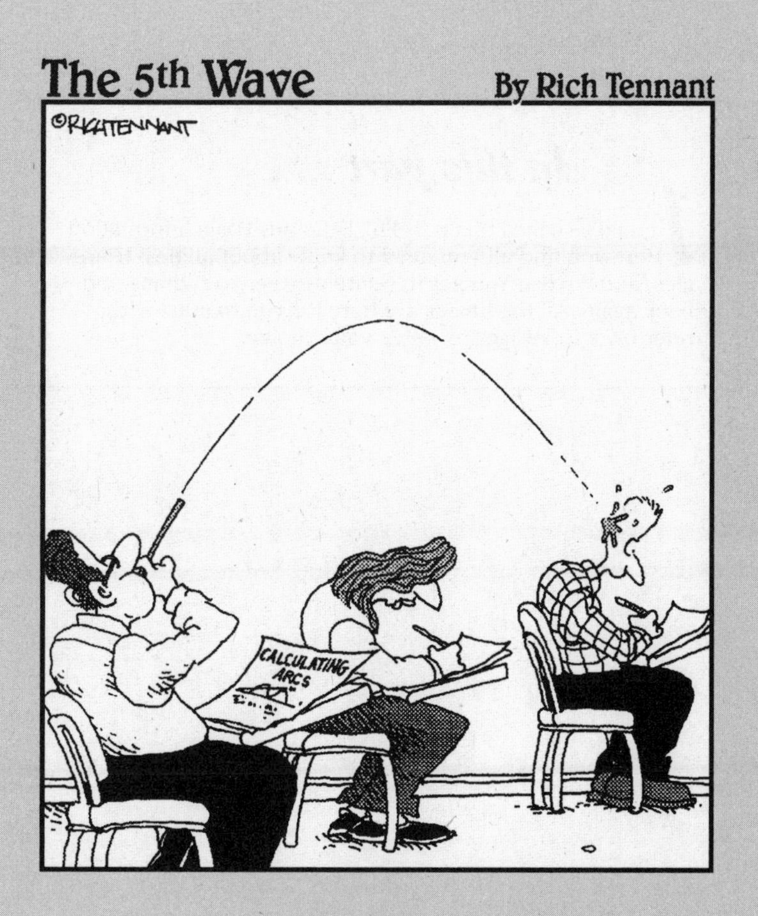

In this part . . .

I could've called this part FBI: Fabulous Basic Information. Here you find all you need to know about angles, triangles, and circles. You get to relate degrees to radians and back again. All the basics are here for you to start with, refer back to, or ignore — it's your choice.

Chapter 1

Tackling Technical Trig

In This Chapter

- Acquainting yourself with angles
- Identifying angles in triangles
- Taking apart circles

Angles are what trigonometry is all about. This is where it all started, way back when. Early astronomers needed a measure to tell something meaningful about the sun and moon and stars and their relationship between man standing on the earth or how they were positioned in relation to one another. Angles are the input values for the trig functions.

This chapter gives you background on how angles are measured, how they are named, and how they relate to one another in two familiar figures, including the triangle and circle. A lot of this material is terminology. The words describe things very specific, but this is a good thing, because they're consistent in trigonometry and other mathematics.

Getting Angles Labeled by Size

An angle is formed where two *rays* (straight objects with an endpoint that go on forever in one direction) have a common endpoint. This endpoint is called the *vertex.* An angle can also be formed when two segments or lines intersect. But, technically, even if it's formed by two segments, those two segments can be extended into rays to describe the angle. Angle measure is sort of *how far apart* the two sides are. The measurement system is unique to these shapes.

Angles can be classified by their size. The measures given here are all in terms of degrees. *Radian measures* (measures of angles that use multiples of π and relationships to the circumference) are covered in Chapter 4, so you can refer to that chapter when needed.

- **Acute angle:** An angle measuring less than 90 degrees.
- **Right angle:** An angle measuring exactly 90 degrees; the two sides are perpendicular.
- **Obtuse angle:** An angle measuring greater than 90 degrees and less than 180 degrees
- **Straight angle:** An angle measuring exactly 180 degrees.

Q. Is an angle measuring 47 degrees acute, right, obtuse, or straight?

A. An angle measuring 47 degrees is acute.

Q. Is an angle measuring 163 degrees acute, right, obtuse, or straight?

A. An angle measuring 163 degrees is obtuse.

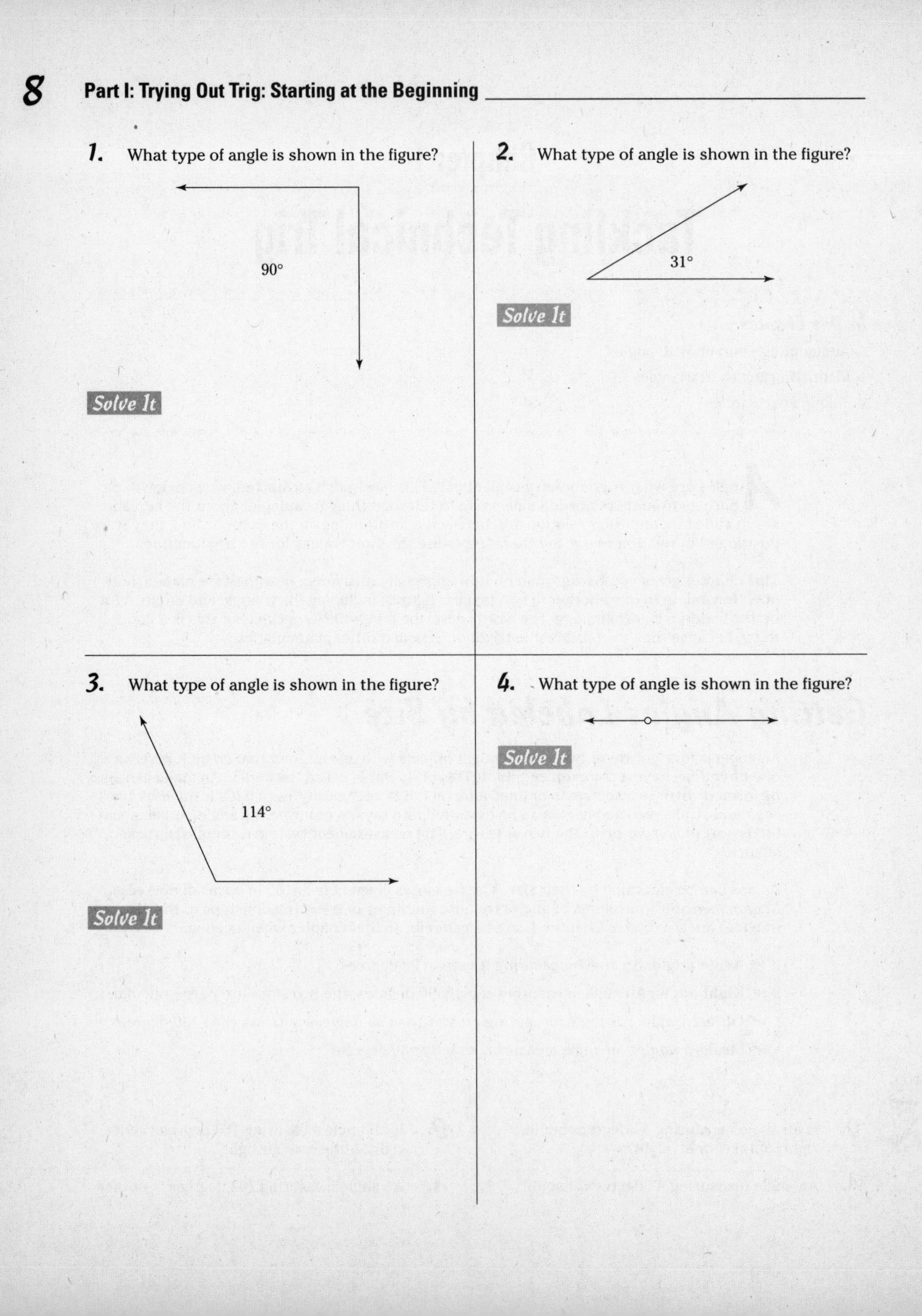

1. What type of angle is shown in the figure?

Solve It

2. What type of angle is shown in the figure?

Solve It

3. What type of angle is shown in the figure?

Solve It

4. What type of angle is shown in the figure?

Solve It

Naming Angles Where Lines Intersect

When two lines cross one another, four angles are formed, and there's something special about the pairs of angles that can be identified there. Look at Figure 1-1. The two lines have intersected, and I've named the angles by putting Greek letters inside them to identify the angles.

Figure 1-1: Two intersecting lines form four angles.

λ β ω θ

The angles that are opposite one another, when two lines intersect, are called *vertical angles.* The special thing they have in common, besides the lines they share, is that their measures are the same, too. There are two pairs of vertical angles in Figure 1-1. Angles β and ω are vertical. So are angles λ and θ.

The other special angles that are formed are pairs of *supplementary angles.* Two angles are supplementary when their sum is 180 degrees. The supplementary angles in Figure 1-1 are those that lie along the same straight line with a shared ray between them. The pairs of supplementary angles are: λ and ω, ω and θ, θ and β, and β and λ.

Q. If one angle in a pair of supplementary angles measures 80 degrees, what does the other angle measure?

A. The other measures 180 – 80 = 100 degrees.

5. Give the measure of the angles that are *supplementary* to the angle shown in the figure.

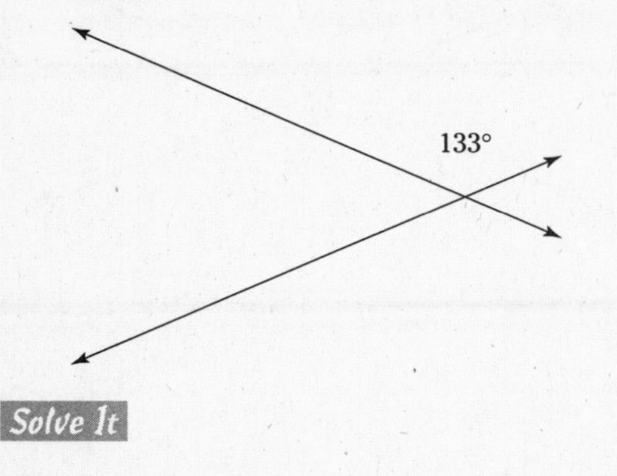

6. Give the measure of the angle that is *vertical* to the angle shown in the figure.

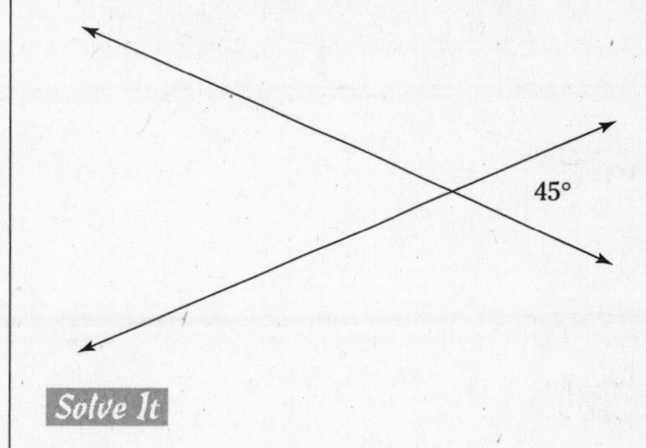

Writing Angle Names Correctly

An angle can be identified in several different ways:

- Use the letter labeling the point that's the *vertex* of the angle. Points are labeled with capital letters.
- Use three letters that label points — one on one ray of the angle, then the vertex, and the last on the other ray.
- Use a letter or number in the inside of the angle. Usually, the letters used are Greek or lowercase.

Q. Give all the different names that can be used to identify the angle shown in the figure.

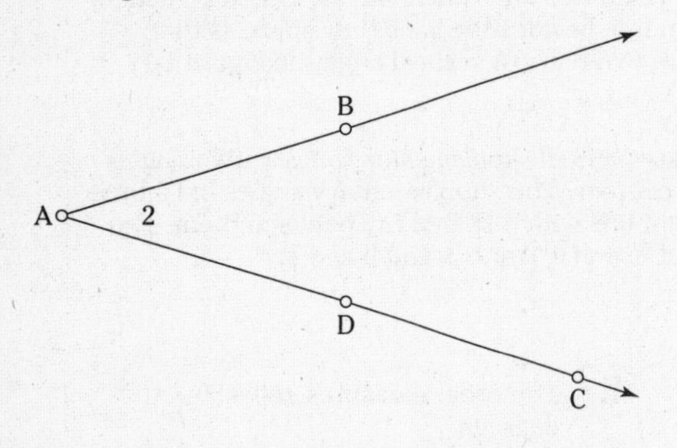

A. The names for this angle are:

- Angle *A* (just using the label for the vertex)
- Angle *BAD* (using *B* on the top ray, the vertex, and *D* on the bottom ray)
- Angle *BAC* (using *B* on the top ray, the vertex, and *C* on the bottom ray)
- Angle *DAB* (using *D* on the bottom ray, the vertex, and *B* on the top ray)
- Angle *CAB* (using *C* on the bottom ray, the vertex, and *B* on the top ray)
- Angle *2* (using the number inside the angle)

7. Find all the names for the angle shown in the figure.

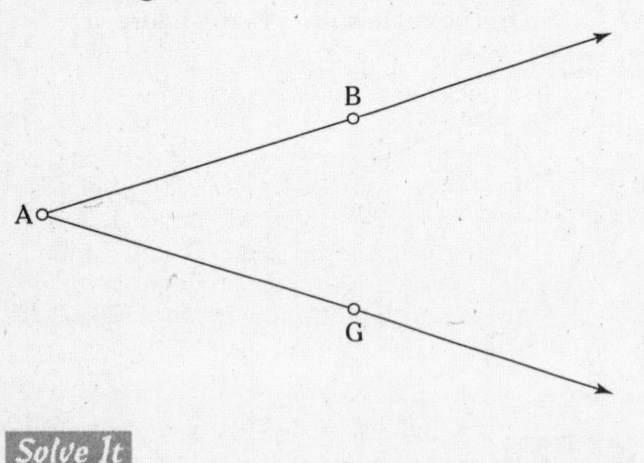

Solve It

8. Find all the names for the angle that's vertical to the angle *POT* in the figure.

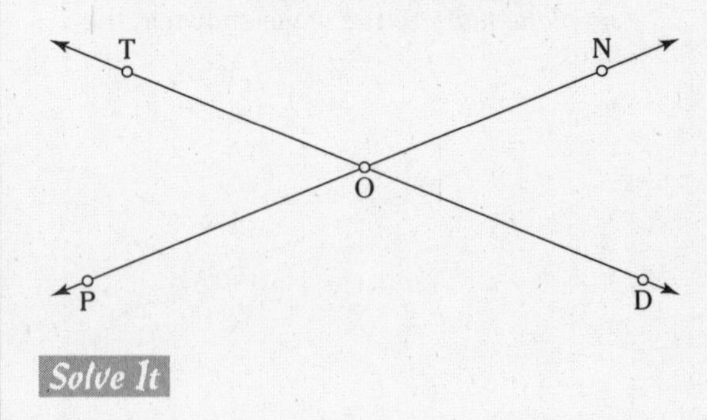

Solve It

Finding Missing Angle Measures in Triangles

Triangles are probably one of the most familiar forms in geometry and trigonometry. They're studied and restudied and gone over for the minutest of details. One thing that stands out, is always true, and is often used, is the fact that the sum of the measures of the angles of any triangle is 180 degrees. It's always that sum — never more, never less. This is a good thing. It allows you to find missing measures — when they go missing — for angles in a triangle.

Q. If the measures of two of the angles of a triangle are 16 degrees and 47 degrees, what is the measure of the third angle?

A. To solve this, add 16 + 47 = 63. Then subtract 180 – 63 = 117 degrees.

Q. An *equilateral triangle* has three equal sides and three equal angles. If you draw a segment from the vertex of an equilateral triangle perpendicular to the opposite side, then what are the measures of the angles in the two new triangles formed? Look at the figure to help you visualize this.

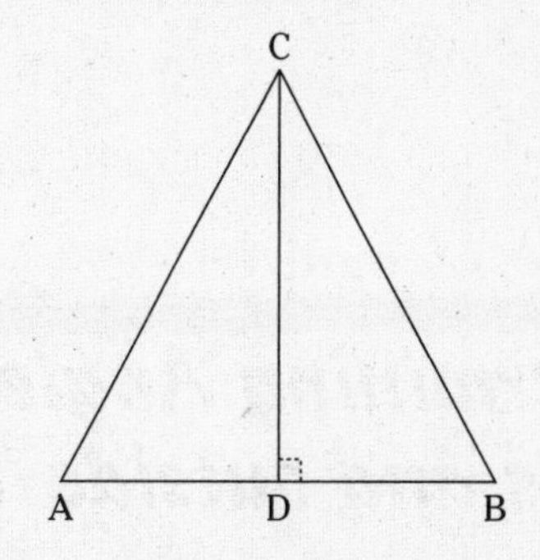

A. Because the triangle is equilateral, the angles must each be 60 degrees, because $3 \times 60 = 180$. That means that angles *A* and *B* are each 60 degrees. If the segment *CD* is perpendicular to the bottom of the triangle, *AB,* then angle *ADC* and angle *BDC* must each measure 90 degrees. What about the two top angles? Because angle *A* is 60 degrees and angle *ADC* is 90 degrees, and because 60 + 90 = 150, that leaves 180 – 150 = 30 degrees for angle *ACD.* The same goes for angle *BCD.*

9. Triangle *SIR* is isosceles. An *isosceles triangle* has two sides that are equal; the angles opposite those sides are also equal. If the vertex angle, *I*, measures 140 degrees, what do the other two angles measure?

10. A triangle has angles that measure n degrees, $n + 20$ degrees, and $3n - 15$ degrees. What are their measures?

Solve It

Determining Angle Measures along Lines and outside Triangles

Angles can be all over the place and arbitrary, or they can behave and be predictable. Two of the situations in the predictable category are those where a transversal cuts through two parallel lines (a *transversal* is another line cutting through both lines), and where a side of a triangle is extended to form an exterior angle.

When a transversal cuts through two parallel lines, the acute angles formed are all equal and the obtuse angles formed are all equal (unless the transversal is perpendicular to the line — in that case, they're all right angles). In Figure 1-2, on the left, you can see how creating acute and obtuse angles comes about. Also, the acute and obtuse angles are supplementary to one another.

An *exterior angle* of a triangle is an angle that's formed when one side of the triangle is extended. The exterior angle is supplementary to the interior angle it's adjacent to. Also, the exterior angle's measure is equal to the sum of the two nonadjacent interior angles.

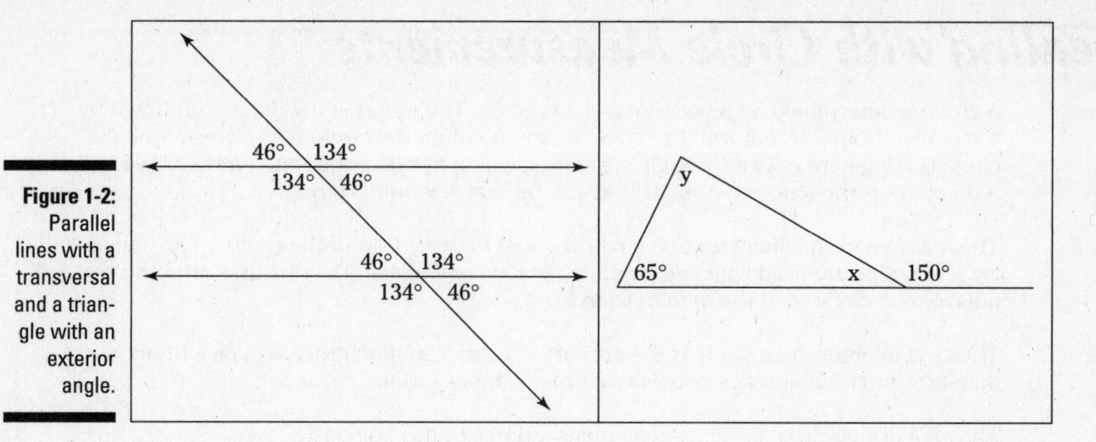

Figure 1-2: Parallel lines with a transversal and a triangle with an exterior angle.

Q. In Figure 1-2, on the right, what are the measures of angles *x* and *y?*

A. The angle *x* is supplementary to an angle of 150 degrees, so its measure is 180 – 150 = 30 degrees. The measure of angle *y* plus the 65-degree angle must equal 150 degrees (the exterior angle's measure). Subtract 150 – 65 = 85. So angle *y* is 85 degrees. To check this, add up the measures of the interior angles: 65 + 85 + 30 = 180. This is the sum of the measures of the angles of any triangle.

11. Find the measures of the acute and obtuse angles formed when a transversal cuts through two parallel lines if the obtuse angles are three times as large as the acute angles.

Solve It

12. Find the measures of the four angles shown in the figure, if one is two times the size of the smallest angle, one is 10 degrees less than five times the smallest angle, and the last is 10 degrees larger than the smallest angle.

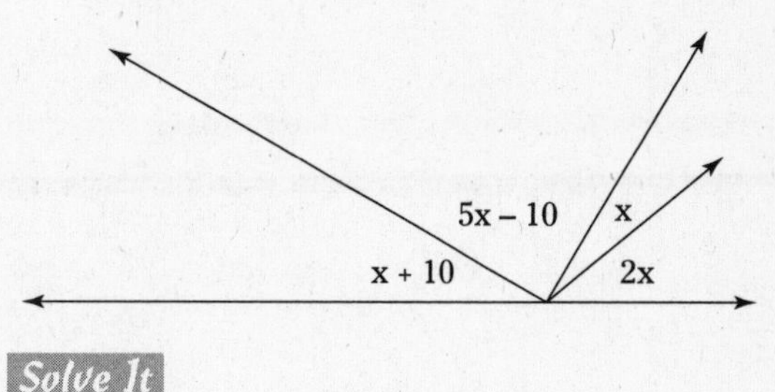

Solve It

Dealing with Circle Measurements

A circle is determined by its center and its radius. The *radius* is the distance, shown by a segment, from the center of the circle to any point on the circle. The *diameter* of a circle is a segment drawn through the center, which has its endpoints on the circle. A diameter is the longest segment that can be drawn within a circle.

The measure of the diameter of a circle is equal to twice that of the radius. The diameter and radius are used when determining the *circumference* (the distance around the outside of a circle) and the *area* of a circle.

The circumference of a circle is $C = \pi d$ or $C = 2\pi r$. Circumference equals π times diameter, or circumference equals two times π times radius.

The area of a circle is $A = \pi r^2$. Area equals π times radius squared.

Q. If a circle has a diameter of 30 inches, find its radius, circumference, and area.

A. If the diameter is 30 inches, then the radius is half that, or 15 inches. The circumference is equal to π times the diameter, so $C = \pi(30) = 30\pi \approx 94.2$ inches. (The squiggly equal sign is a way of showing that the measure is "about" that much, not exactly equal to that much.) The approximation was obtained letting $\pi \approx 3.14$. And the area is equal to $A = \pi(15)^2 = 225\pi \approx 706.5$ square inches.

13. Find the radius, circumference, and area of a circle that has a diameter of $2\sqrt{3}$ yards.

Solve It

14. Find the diameter, radius, and area of a circle that has a circumference of 18π centimeters.

Tuning In with the Right Chord

A *chord* is a segment that's drawn from one point on a circle to another point on the same circle. The longest chord of a circle is its diameter. The two endpoints of a chord divide a circle into two arcs — the major arc and the minor arc. The major arc is, of course, the larger of the two. A circle has a total of 360 degrees, so the sum of those two arcs must equal 360.

Q. The chord *AB*, shown in the figure, divides the circle into two arcs, one of which is 100 degrees greater than the other. What is the measure of the major arc?

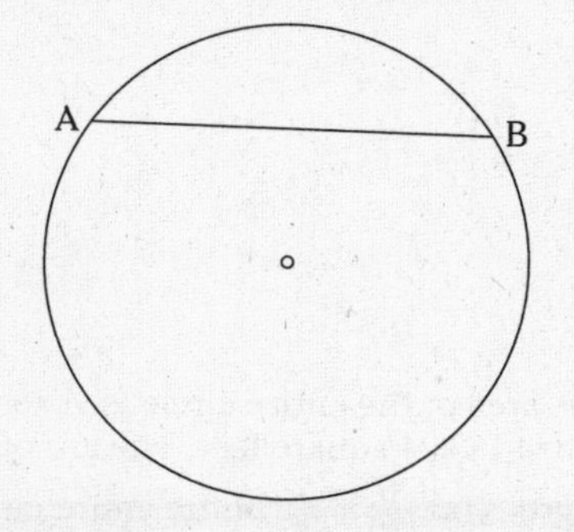

A. Let the measure of the minor arc be x. Then the larger arc is 100 greater than that, or $x + 100$. The sum of the two is 360. Write that as $x + x + 100 = 360$. This simplifies to $2x + 100 = 360$. Subtract 100 from each side to get $2x = 260$. Divide by 2, and $x = 130$. This is the measure of the minor arc. Add 100 to that, and the major arc measures 230 degrees.

15. A chord divides a circle into two arcs, one of which is 15 degrees less than 14 times the other. What are the measures of the two arcs?

Solve It

16. Three chords are drawn in a circle to form a triangle, as shown in the figure. One of the chords is drawn through the center of the circle. If the minor arc determined by the shortest chord is 60 degrees, what are the measures of the other two arcs determined by the vertices of the triangle?

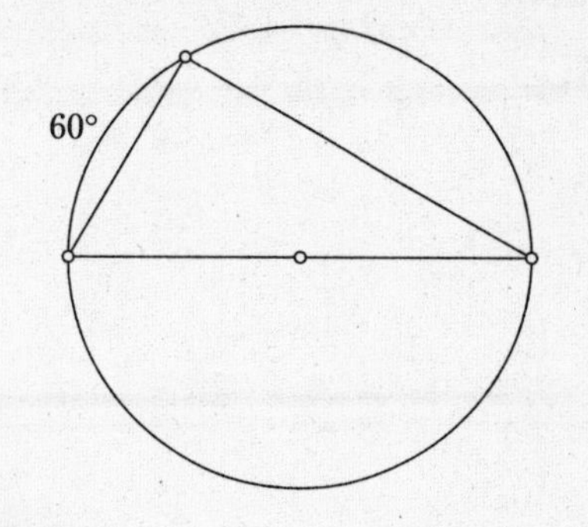

Solve It

Sectioning Off Sectors of Circles

A sector of a circle is a wedge or slice of it. Look at Figure 1-3, showing a sector of a circle that has an arc that measures 70 degrees.

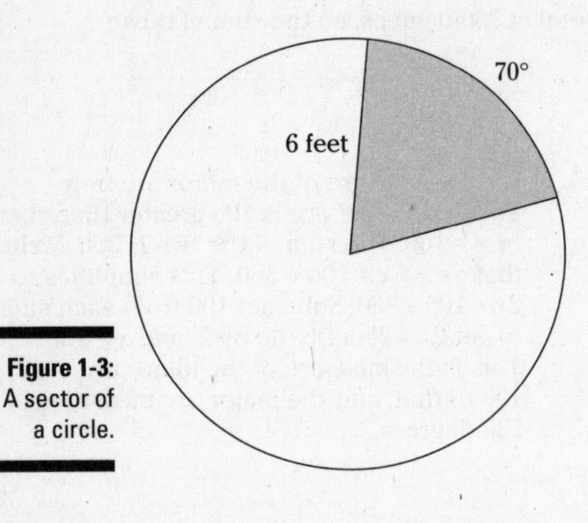

Figure 1-3: A sector of a circle.

Q. Find the area of the sector, shown in Figure 1-3, that has an arc of 70 degrees and a radius of 6 feet.

A. The area of the entire circle is $A = \pi(6)^2 = 36\pi \approx 113.04$ square feet. The arc of 70 degrees is $\frac{70}{360} = \frac{7}{36}$ of the entire circle. So multiply the area of the entire circle by that fraction to get the area of the sector. $36\pi\left(\frac{7}{36}\right) = \pi(7) \approx 21.98$ square inches.

17. Find the area of the sector of a circle that has an arc measuring 120 degrees and a radius of 2.4 meters.

18. A pizza is being divided into three unequal slices (sectors). The largest slice has an arc measuring 1 less than three times that of the smallest slice's arc, and the middle-sized piece has an arc that's 1 more than twice the smallest slice's arc. If this is an 18-inch pizza, what is the area of each of the pieces?

Answers to Problems on Tackling Technical Trig

The following are the solutions to the practice problems presented earlier in this chapter.

1 What type of angle is shown in the figure? **Right angle.**

The angle shown in the figure is a right angle, because it measures exactly 90 degrees.

2 What type of angle is shown in the figure? **Acute angle.**

The angle shown in the figure is an acute angle, because 31 is between 0 and 90 degrees.

3 What type of angle is shown in the figure? **Obtuse angle.**

The angle shown in the figure is an obtuse angle, because 114 is between 90 and 180 degrees.

4 What type of angle is shown in the figure? **Straight angle.**

The angle shown in the figure is a straight angle, because it measures exactly 180 degrees.

5 Give the measure of the angles that are *supplementary* to the angle shown in the figure. **47 degrees.**

In the figure, the measure of the angles that are *supplementary* to the 133-degree angle is 47 degrees, because 180 – 133 = 47.

6 Give the measure of the angle that is *vertical* to the angle shown in the figure. **45 degrees.**

In the figure, the measure of the angle that is *vertical* to the 45-degree angle is also 45 degrees, because vertical angles always have the same measure.

7 Find all the names for the angle shown in the figure. ***A, BAG,* and *GAB.***

8 Find all the names for the angle that's vertical to the angle *POT* in the figure. ***NOD* and *DON.***

In the figure, you can't use the letter *O* to name an angle. This is a case where just using the letter for the point at the vertex doesn't give enough information to identify which angle you're talking about.

9 Triangle *SIR* is isosceles. An *isosceles triangle* has two sides that are equal; the angles opposite those sides are also equal. If the vertex angle, *I,* measures 140 degrees, what do the other two angles measure? **20 degrees each.**

If the vertex angle, *I,* measures 140 degrees, the other two angles have to be the angles that are equal. The reason for this is that, if the 140-degree angle were one of the pair of equal angles, the sum of it and its pair would be 280 degrees, which is already too much for the sum of the angles in a triangle. So, to find the measure of the two equal angles, first subtract 180 – 140 = 40. That leaves a total of 40 degrees for the two equal angles; they're 20 degrees each.

10 A triangle has angles that measure n degrees, $n + 20$ degrees, and $3n - 15$ degrees. What are their measures? **35 degrees, 55 degrees, and 90 degrees.**

The angles have to add up to 180 degrees: $n + (n + 20) + (3n - 15) = 180$. Simplifying on the left, $5n + 5 = 180$. Subtract 5 from each side to get $5n = 175$. Divide each side by 5, and $n = 35$. And $n + 20 = 55$. Lastly, $3n - 15 = 90$. Adding the three angles: $35 + 55 + 90 = 180$.

11 Find the measures of the acute and obtuse angles formed when a transversal cuts through two parallel lines if the obtuse angles are three times as large as the acute angles. **45 degrees and 135 degrees.**

Let the sum of the acute and obtuse angles be 180. This is true because the angles are supplementary. Let the acute angle's measure be x. Then the obtuse angle measures $3x$. Adding, $x + 3x = 180$, $4x = 180$. Dividing by 4, $x = 45$. If the acute angles are 45 degrees, then the obtuse angles are three times that, or 135 degrees.

12 Find the measures of the four angles shown in the figure, if one is two times the size of the smallest angle, one is 10 degrees less than five times the smallest angle, and the last is 10 degrees larger than the smallest angle. **20 degrees, 40 degrees, 90 degrees, and 30 degrees.**

The sum of the four angles is 180 degrees — the measure of a straight angle. Let the smallest measure be x degrees. Then the others are $2x$, $5x - 10$, and $x + 10$. Adding them, $x + 2x + 5x - 10 + x + 10 = 180$. Simplifying on the left, $9x = 180$. Dividing by 9, $x = 20$. The angles are: 20, 40, 90, and 30 degrees.

13 Find the radius, circumference, and area of a circle that has a diameter of $2\sqrt{3}$ yards. $\frac{1}{2}(2\sqrt{3}) = \sqrt{3}$**,** $C = \pi(2\sqrt{3}) = 2\pi\sqrt{3}$ **yards, and** 3π **square yards.**

The radius is half the diameter, so half of $2\sqrt{3}$ is $\frac{1}{2}(2\sqrt{3}) = \sqrt{3}$. The circumference is π times the diameter, so $C = \pi(2\sqrt{3}) = 2\pi\sqrt{3}$ yards. The area is π times the square of the radius, so $A = \pi(\sqrt{3})^2 = \pi(3) = 3\pi$ square yards.

14 Find the diameter, radius, and area of a circle that has a circumference of 18π centimeters. **18 centimeters, 9 centimeters, and** 81π **square centimeters.**

The circumference is π times the diameter, so the diameter must be 18 centimeters. That means that the radius is half that, or 9 centimeters. The area is π times the square of the radius or $A = \pi(9)^2 = \pi(81) = 81\pi$ square centimeters.

15 A chord divides a circle into two arcs, one of which is 15 degrees less than 14 times the other. What are the measures of the two arcs? **25 degrees and 335 degrees.**

Start by finding the measures of the two arcs. Let one arc measure x degrees. Then the other measures $14x - 15$ degrees. Their sum is 360 degrees. So $x + 14x - 15 = 360$. Simplify on the left to get $15x - 15 = 360$. $15x = 375$. Dividing by 15, $x = 25$. The minor arc is 25 degrees, and the major arc is $14(25) - 15 = 335$ degrees.

16 Three chords are drawn in a circle to form a triangle, as shown in the figure. One of the chords is drawn through the center of the circle. If the minor arc determined by the shortest chord is 60 degrees, what are the measures of the other two arcs determined by the vertices of the triangle? **180 degrees and 120 degrees.**

The diameter divides the circle into two equal arcs, so they're each 180 degrees. That leaves 180 degrees for the top half. Subtract 180 – 60, and the other arc on the top is 120 degrees.

17 Find the area of the sector of a circle that has an arc measuring 120 degrees and a radius of 2.4 meters. 1.92π **square meters.**

The sector is $\frac{120}{360} = \frac{1}{3}$ of the entire circle. Multiply that times the area of the entire circle, which is found by multiplying π times the square of the radius: $\frac{1}{3}\pi(2.4)^2 = \frac{5.76}{3}\pi = 1.92\pi$ square meters.

18 A pizza is being divided into three unequal slices (sectors). The largest slice has an arc measuring 1 less than three times that of the smallest slice's arc, and the middle-sized piece has an arc that's 1 more than twice the smallest slice's arc. If this is an 18-inch pizza, what is the area of each of the pieces? $\approx$ **42.39 square inches,** $\approx$ **126.46 square inches, and** $\approx$ **85.49 square inches.**

Let x represent the measure of the arc cut by the smallest piece. Then the other two arcs are $3x - 1$ and $2x + 1$. Add all three together to get $x + 3x - 1 + 2x + 1 = 360$. Simplifying on the left, $6x = 360$. Dividing by 6, $x = 60$. The other two measures are then 179 and 121. The three different areas can be found by multiplying their fraction of the pizza by the area of the whole pizza, which is determined by multiplying π times the square of the radius. An 18-inch pizza has a radius of 9 inches.

$$\frac{60}{360}\pi(9)^2 = \frac{4860}{360}\pi \approx 42.39 \text{ square inches}$$

$$\frac{179}{360}\pi(9)^2 = \frac{14499}{360}\pi \approx 126.46 \text{ square inches}$$

$$\frac{121}{360}\pi(9)^2 = \frac{9801}{360}\pi \approx 85.49 \text{ square inches}$$

Chapter 2

Getting Acquainted with the Graph

In This Chapter

- Putting points where they belong
- Measuring and minding the middle
- Conferring with Pythagoras
- Circling about and becoming well-rounded

Graphing points and figures is an important technique in anything mathematical, and it's especially important in trigonometry. Not only is the coordinate plane used to describe how figures are interacting, but it's also used to define the six trig functions in all their glory. The values of the trig functions are dependent upon where the inputs are in the graph. And, of course, there are the graphs of the functions, which I cover in detail in Part IV.

The important things to look for in the graphs of the trig functions and other curves is where the graph is and when it's there. The signs of the coordinates bear watching. The quadrant that the curve lies in is important. These are all covered in this chapter.

Plotting Points

The traditional setup for graphing in mathematics is the *coordinate plane,* which is defined by two perpendicular lines intersecting at the center *(origin)* and uses the *Cartesian coordinates.* The coordinates are the numbers representing the placement of the points. These numbers are in *ordered pairs* — they're written in parentheses, separated by a comma, and the order they're in matters. The points are in the form (x,y) where the first coordinate, the x, tells you how far to the right or left the point is in terms of the origin. The second coordinate, the y, tells you how far up or down the point is from the origin. Positive coordinates indicate a move to the right or up. Negative coordinates indicate to the left or down. In Figure 2-1, I've graphed the points (3,2), (–4,3), (–5,–2), (6,–3), (0,5), and (–2,0). Notice that the points that have a 0 as one of the coordinates lie on one of the axes.

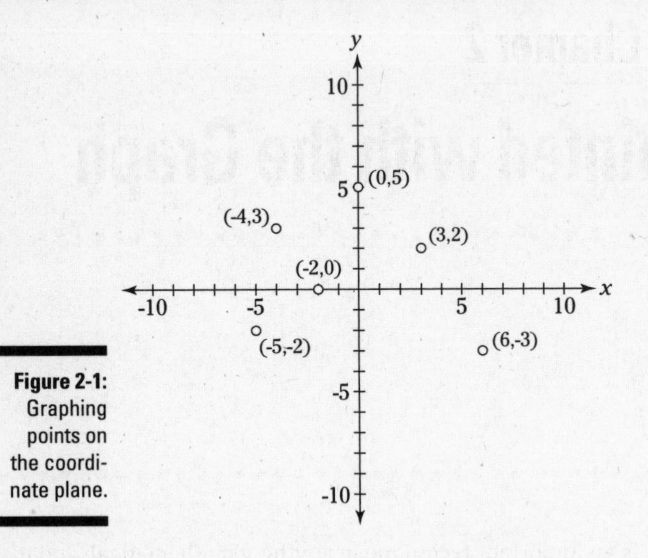

Figure 2-1: Graphing points on the coordinate plane.

Q. Refer to Figure 2-1. Give the coordinates of the points that are on the opposite side of the *y*-axis from the points that are graphed there.

A. The point opposite (3,2) is (-3,2); the point opposite (–4,3) is (4,3); the point opposite (–5,–2) is (5,–2); the point opposite (6,–3) is (–6,–3); and the point opposite (–2,0) is (2,0). Since the point (0,5) is on the *y*-axis, it has no opposite.

Q. Give the coordinates of the points that are on the opposite side of the *x*-axis from the points (–2,3), (4,–7) and (–6,–1).

A. The point opposite (–2,3) is (–2,-3); the point opposite (4,–7) is (4,7); and the point opposite (–6,–1) is (–6,1).

1. Plot the following points on the coordinate plane: (2,–3), (4,2), (–5,3), (–1,–4).

Solve It

2. Plot the following points on the coordinate plane: (3,0), (–4,0), (0,5), (0,–2).

Solve It

Identifying Points by Quadrant

The coordinate plane is divided by the intersecting axes into four *quadrants*. These quadrants are traditionally given the labels I, II, III, and IV. The Roman numerals start in the upper-right-hand corner and go counterclockwise from there. In each quadrant, the points that lie there have some common properties. The quadrants are used when determining values for the trig functions. Angles are graphed on the coordinate plane, and their positions with respect to the quadrants is important. Figure 2-2 shows you the names of the different quadrants.

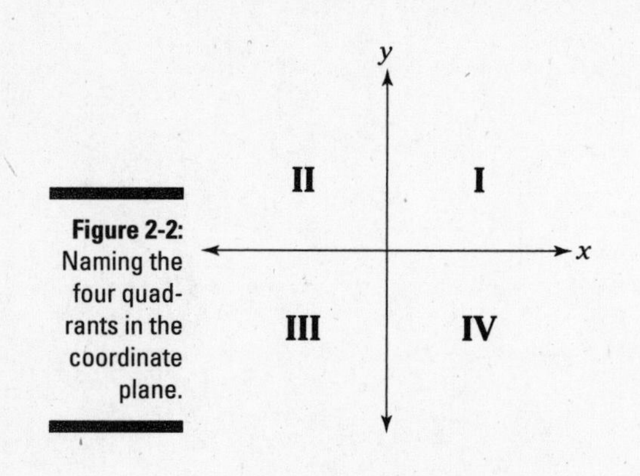

Figure 2-2: Naming the four quadrants in the coordinate plane.

The common characteristics of the points in the quadrants are as follows:

Quadrant I: Both the x and y coordinates are positive.

Quadrant II: The x coordinate is negative, and the y coordinate is positive.

Quadrant III: Both the x and y coordinates are negative.

Quadrant IV: The x coordinate is positive, and the y coordinate is negative.

The points that lie on axes are not considered to be in any quadrant.

Q. Name the quadrant that the point (–6,2) lies in.

A. Quadrant II.

Q. Name the quadrant that the point (–4,–8) lies in.

A. Quadrant III.

3. Give the coordinates of three different points that lie in Quadrant III.

Solve It

4. Give the coordinates of three different points that do not lie in any quadrant.

Solve It

Working with Pythagoras

The Pythagorean theorem is probably one of the most easily recognized rules or equations in all of mathematics. The equation itself is easy to remember. And even if you can't put it into words exactly, you pretty much know how it works.

Pythagoras recognized (as did others, but he got the credit for it) that the lengths of the sides of any right triangle have a special relationship to one another. If you square the lengths of all the sides and add up the two smaller squares, that's always equal to the larger square. Here, in math-talk, is the Pythagorean theorem:

> If the lengths of the two *legs* (shorter sides) of a right triangle are a and b, and if the length of the *hypotenuse* (the longest side, opposite the right angle) is c, then $a^2 + b^2 = c^2$.

Q. Find the length of side a in a right triangle if the other leg, b, measures 32 feet and the hypotenuse, c, measures 40 feet.

A. Replace the a and c in the Pythagorean theorem with the two given lengths: $a^2 + 32^2 = 40^2$. Square the values, and solve for b.

$$a^2 + 1024 = 1600$$
$$a^2 = 1600 - 1024 = 576$$
$$a = \sqrt{576} = 24$$

So side a is 24 feet long.

5. Find the length of the hypotenuse of a right triangle if $a = 7$ yards and $b = 24$ yards.

Solve It

6. Find the length of leg b in a right triangle with $a = 18$ feet and $c = 30$ feet.

7. How long is leg a in the right triangle with $b = 24$ meters and $c = 26$ meters?

Solve It

8. A right triangle has one leg, a, that measures 9 inches. The other leg, b, is 1 inch shorter than the hypotenuse. How long are that leg and the hypotenuse?

Keeping Your Distance

The distance formula for finding the length of the segment between points on the coordinate plane finds the horizontal distance between the points by finding the difference between the *x*-coordinates, finds the vertical distance between the points by finding the difference between the *y*-coordinates, and then actually uses the Pythagorean theorem to solve for the slant difference between them. The formula $d = \sqrt{(x_1 - x_2)^2 + (y_1 - y_2)^2}$ has a *d* for distance instead of a *c* for the length of the hypotenuse of a right triangle, but the arithmetic is still the same. The formula in this form is just more convenient for doing the computations.

Q. Find the distance between the points (–3,5) and (6,–7).

A. First plug the coordinates of the points into the formula: $d = \sqrt{(-3-6)^2 + (5-(-7))^2}$.

Then find the differences, square the results, and add the squares together.

Finally, find the square root of the sum. That's the distance between the points in units on the coordinate plane.

$$= \sqrt{(-9)^2 + (12)^2}$$
$$= \sqrt{81 + 144} = \sqrt{225}$$
$$= 15$$

9. Find the distance between the points (3,–8) and (–5,–2).

Solve It

10. Find the distance between the points (0,5) and (12,0).

Solve It

11. Find the distance between the points (–4,11) and (5,–2).

Solve It

12. Find the distance between the points (6,3) and (6,–7).

Finding Midpoints of Segments

The midpoint of a segment that's graphed on the coordinate plane can be found by simply averaging the *x* and *y* coordinates. Finding the midpoint is important if you're working with circles and need a center or if you're working with perpendicular bisectors and need the bisector. Here's the formula for the midpoint of a segment with endpoints (x_1, y_2) and (x_2, y_2):

$$M = \left(\frac{x_1 + x_2}{2}, \frac{y_1 + y_2}{2}\right)$$

Q. Find the midpoint of the segment with endpoints (–3,4) and (7,10).

A. Using the formula,

$$M = \left(\frac{-3+7}{2}, \frac{4+10}{2}\right) = \left(\frac{4}{2}, \frac{14}{2}\right) = (2, 7).$$

13. Find the midpoint of the segment with endpoints (4,–6) and (–8,10).

Solve It

14. Find the midpoint of the segment with endpoints (–9,–3) and (15,–5). Then find the midpoints of the two segments formed. (This, in effect, divides the segment into four equal parts.)

Dealing with Slippery Slopes

The slope of a line or segment tells you something about the character of that line or segment. The slope is a number — either positive or negative — that tells you whether a line or segment is rising as you read from left to right or falling as you read from left to right. The positive slopes rise; the negative slopes fall. The rest of the information has to do with the steepness of the line or segment.

A line with a slope of 1 or –1 is diagonal going upward or downward and, if it goes through the origin, it bisects the two quadrants it goes through. Slopes that are greater than 1 or smaller than –1 (such as slopes of 7, –5, 4.2, –9.3, and so on) indicate lines that are steep. They're steeper than those with the benchmark slope of 1. Slopes that are proper fractions — the values are between –1 and 1 — are for lines that are fairly flat. The closer the value of the slope is to 0, the closer the line or segment is to being horizontal.

Slopes of lines can be determined by choosing any two points on the line and putting them into the slope formula. If you have two points on a line, (x_1, y_1) and (x_2, y_2), then the slope is found by using the slope formula, $m = \frac{y_1 - y_2}{x_1 - x_2}$.

Slopes of lines can also be determined if you have the equation of the line in the *slope-intercept form.* This form is $y = mx + b$, where m is the slope of the line.

EXAMPLE

Q. Find the slope of the line with the equation $y = -3x + 5$ using the slope-intercept form and then by finding two points on the line and using the formula for slope.

A. From the slope-intercept form, the slope of this line is –3. Using the slope formula, two points have to be determined first. I choose to let $x = 1$, giving me $y = 2$. Then I choose $x = -2$, making $y = 11$. These two choices are absolutely random. Any points will work. Now, I insert these two points, (1,2) and (–2,11), into the formula, $m = \frac{2-11}{1-(-2)} = \frac{-9}{3} = -3$.

15. Find the slope of the line that goes through (8,2) and (–3,–9).

Solve It

16. Find the slope of the line that goes through (4,–1) and (5,–1).

Solve It

17. Find the slope of the line $y = 4x - 1$ using the slope-intercept form. Then check your answer by choosing two points on the line and using the slope formula.

18. Find the slope of the line $3x - 2y = 6$ using the slope-intercept form. (***Hint:*** First solve for y to put the equation into the proper form.)

Writing Equations of Circles

A circle plays a big part in trigonometry, because it's used when formulating measures of angles and function values for the trig functions. The *unit circle,* a circle usually centered at the origin and always with a radius of 1 unit, is used the most, but the rest of the circles need to be discussed and worked with, too. Here are the equations of circles you'll need:

The equation of any circle with its center at the origin is $x^2 + y^2 = r^2$, where r is the radius of the circle.

The equation of the *unit circle* centered at the origin is $x^2 + y^2 = 1$.

The equation of any circle is $(x - h)^2 + (y - k)^2 = r^2$, where (h,k) are the coordinates of the center of the circle and r is the radius.

Q. Identify the center and radius of the circle whose equation is $(x - 5)^2 + (y + 9)^2 = 100$.

A. The center is at the point (5,–9). Notice that there's a $y + 9$ in the second parentheses. Think of that as being $y - (-9)$ to fit the form. The radius is 10, because 100 is the square of 10.

Q. Write the equation of the circle whose center is at (–3,0) and that has a radius of 16.

A. The equation is $(x + 3)^2 + y^2 = 256$. Notice that the 0 isn't shown. It isn't necessary here; you assume that the 0 was subtracted and the equation simplified.

19. Identify the center and radius of the circle $(x-3)^2+(y+2)^2=9$.

Solve It

20. Identify the center and radius of the circle $(x+5)^2+(y-6)^2=16$.

Solve It

21. Write the equation of the circle with its center at (3,2) and a radius of 6.

Solve It

22. Write the equation of the circle with its center at (–4,3) and a radius of $\frac{1}{3}$.

Graphing Circles

A circle is determined, uniquely, by its center and radius. Graphing circles is done quickly and accurately by first determining where the center is on the coordinate plane, and then counting off the radius to the left, right, above, and below that center. The circle can then be drawn in using the endpoints of those radii.

Q. Sketch the graph of the circle $(x+1)^2+(y-1)^2=4$.

A. The center is at (–1,1), and the radius is 2 units long. Going up left 2 units from the center (–1,1), the endpoint of the radius is (–3,1). Going right 2 units from the center, the point on the circle is (1,1). Going up 2 units from the center, the endpoint is (–1,3). And downward 2 units from the center is the point (–1,–1). The figure shows a sketch of that circle.

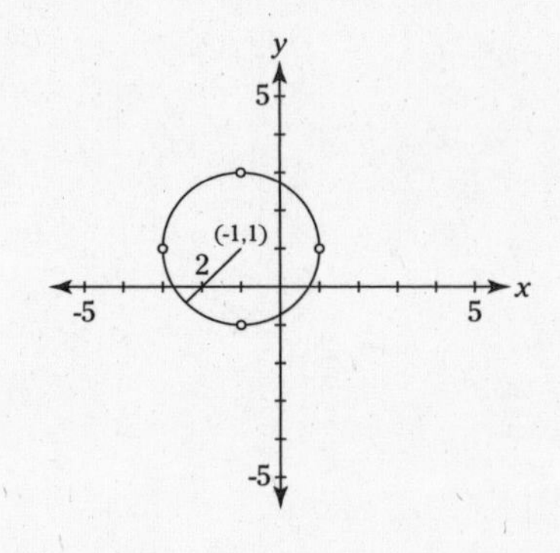

23. Sketch the graph of the circle $(x-3)^2+(y+2)^2=16$.

24. Sketch the graph of the circle $x^2+(y-4)^2=36$.

Answers to Problems on Graphing

The following are the solutions to the practice problems presented earlier in this chapter.

1 Plot the following points on the coordinate plane: (2,–3), (4,2), (–5,3), (–1,-4). **See the following figure.**

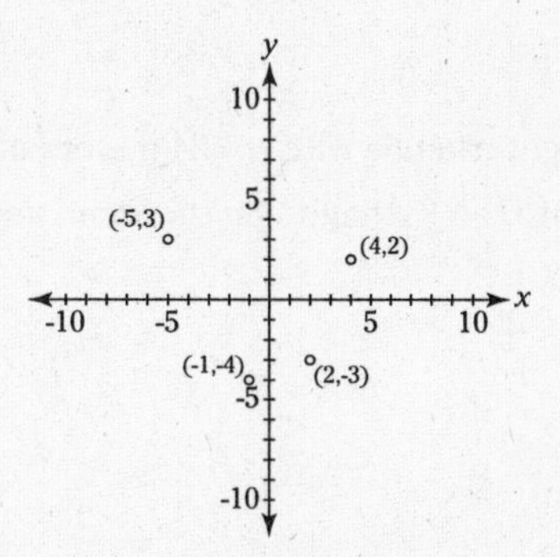

2 Plot the following points on the coordinate plane: (3,0), (–4,0), (0,5), (0,–2). **See the following figure.**

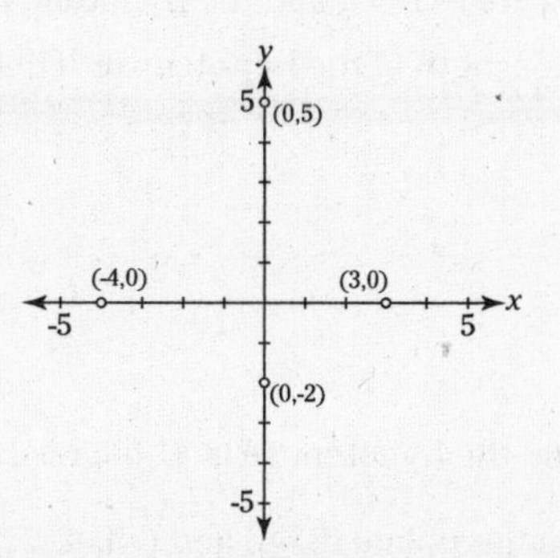

3 Give the coordinates of three different points that lie in quadrant III. **(–3,–2), (–4,–6), and (–1,–2).**

Unless you can read my mind, your answers probably vary. But, in each case, both the x and y coordinates must be negative numbers, and you won't have any zeroes.

4 Give the coordinates of three different points that do not lie in any quadrant. **(0,2), (–3,0), and (5,0).**

Your answers probably vary. But, in each case, at least one of the coordinates will be a zero.

5 Find the length of the hypotenuse of a right triangle if a = 7 yards and b = 24 yards. **25 yards.**

Substitute the numbers into the Pythagorean theorem, and solve for c:

$$7^2 + 24^2 = c^2$$
$$49 + 576 = c^2$$
$$625 = c^2$$
$$25 = c$$

The hypotenuse is 25 yards long.

6 Find the length of leg b in a right triangle with a = 18 feet and c = 30 feet. **24 feet.**

Substitute the numbers into the Pythagorean theorem, and solve for b:

$$18^2 + b^2 = 30^2$$
$$324 + b^2 = 900$$
$$b^2 = 900 - 324$$
$$b^2 = 576$$
$$b = 24$$

The leg b is 24 feet long.

7 How long is leg a in the right triangle with b = 24 meters and c = 26 meters? **10 meters.**

Substitute the numbers into the Pythagorean theorem, and solve for a:

$$a^2 + 24^2 = 26^2$$
$$a^2 + 576 = 676$$
$$a^2 = 676 - 576$$
$$a^2 = 100$$
$$a = 10$$

The leg a is 10 meters long.

8 A right triangle has one leg, a, that measures 9 inches. The other leg, b, is 1 inch shorter than the hypotenuse. How long are that leg and the hypotenuse? **40 inches and 41 inches.**

Let $b = c - 1$, where c is the length of the hypotenuse. Then, substitute the numbers into the Pythagorean theorem:

$$9^2 + (c-1)^2 = c^2$$
$$81 + c^2 - 2c + 1 = c^2$$
$$82 - 2c = 0$$
$$82 = 2c$$
$$41 = c$$

Leg b is 40 inches long, and the hypotenuse is 41 inches long.

9 Find the distance between the points (3,–8) and (–5,–2). **10.**

Substitute the coordinates into the distance formula:

$$d = \sqrt{[3-(-5)]^2 + [-8-(-2)]^2}$$
$$= \sqrt{8^2 + (-6)^2}$$
$$= \sqrt{64 + 36}$$
$$= \sqrt{100} = 10$$

10 Find the distance between the points (0,5) and (12,0). **13.**

Substitute the coordinates into the distance formula:

$$d = \sqrt{(0-12)^2 + (5-0)^2}$$
$$= \sqrt{144 + 25}$$
$$= \sqrt{169} = 13$$

11 Find the distance between the points (–4,11) and (5,–2). ≈ **15.811.**

Substitute the coordinates into the distance formula:

$$d = \sqrt{(-4-5)^2 + [11-(-2)]^2}$$
$$= \sqrt{(-9)^2 + 13^2}$$
$$= \sqrt{81+169}$$
$$= \sqrt{250}$$
$$= \sqrt{25}\sqrt{10} = 5\sqrt{10}$$
$$\approx 15.811$$

12 Find the distance between the points (6,3) and (6,–7). **10.**

Substitute the coordinates into the distance formula:

$$d = \sqrt{(6-6)^2 + [3-(-7)]^2}$$
$$= \sqrt{0+10^2}$$
$$= \sqrt{100} = 10$$

These points are on the vertical line. The distance is just the difference between the two *y* values.

13 Find the midpoint of the segment with endpoints (4,–6) and (–8,10). **(–2,2).**

Substitute into the formula:

$$M = \left(\frac{4+(-8)}{2}, \frac{-6+10}{2}\right)$$
$$= \left(\frac{-4}{2}, \frac{4}{2}\right) = (-2, 2)$$

14 Find the midpoint of the segment with endpoints (–9,–3) and (15,–5). Then find the midpoints of the two segments formed. (This, in effect, divides the segment into four equal parts.) **(3,–4), (–3,$-\frac{7}{2}$), (9,$-\frac{9}{2}$).**

First find the midpoint of the two given points:

$$M = \left(\frac{-9+15}{2}, \frac{-3+(-5)}{2}\right)$$
$$= \left(\frac{6}{2}, \frac{-8}{2}\right) = (3, -4)$$

Then find the midpoint of the segment determined by (–9,–3) and this midpoint:

$$M = \left(\frac{-9+3}{2}, \frac{-3+(-4)}{2}\right)$$
$$= \left(\frac{-6}{2}, \frac{-7}{2}\right) = \left(-3, \frac{-7}{2}\right)$$

Lastly, find the midpoint of the segment determined by (15,–5) and the original midpoint:

$$M = \left(\frac{3+15}{2}, \frac{-4+(-5)}{2}\right)$$
$$= \left(\frac{18}{2}, \frac{-9}{2}\right) = \left(9, \frac{-9}{2}\right)$$

15 Find the slope of the line that goes through (8,2) and (–3,–9). **1.**

Substitute into the formula:

$$m = \frac{2-(-9)}{8-(-3)} = \frac{11}{11} = 1$$

16 Find the slope of the line that goes through (4,–1) and (5,–1). **0.**

Substitute into the formula:

$$m = \frac{-1-(-1)}{4-5} = \frac{0}{-1} = 0$$

This is a horizontal line; the *y* values are the same. The slope of a horizontal line is 0.

17 Find the slope of the line $y = 4x - 1$ using the slope-intercept form. Then check your answer by choosing two points on the line and using the slope formula. **4.**

Using the slope-intercept form, since the coefficient of *x* is 4, the slope is 4. To check this, choose two points on the line. The two points I've chosen (these are random — yours may be different) are (1,3) and (–2,–9).

$$m = \frac{3-(-9)}{1-(-2)} = \frac{12}{3} = 4$$

18 Find the slope of the line $3x - 2y = 6$ using the slope-intercept form. (***Hint:*** First solve for *y* to put the equation into the proper form.) $\mathbf{\frac{3}{2}}$.

First subtract 3*x* from each side of the equation: $-2y = -3x + 6$. Then divide each term by –2 to get $y = \frac{3}{2}x - 3$. The slope is $\frac{3}{2}$.

19 Identify the center and radius of the circle $(x-3)^2 + (y+2)^2 = 9$. **(3,–2), 3.**

20 Identify the center and radius of the circle $(x+5)^2 + (y-6)^2 = 16$. **(–5,6), 4.**

21 Write the equation of the circle with its center at (3,2) and a radius of 6. $\mathbf{(x-3)^2 + (y-2)^2 = 36}$.

22 Write the equation of the circle with its center at (–4,3) and a radius of $\frac{1}{3}$. $\mathbf{(x+4)^2 + (y-3)^2 = \frac{1}{9}}$.

23 Sketch the graph of the circle $(x-3)^2 + (y+2)^2 = 16$. **See the following figure.**

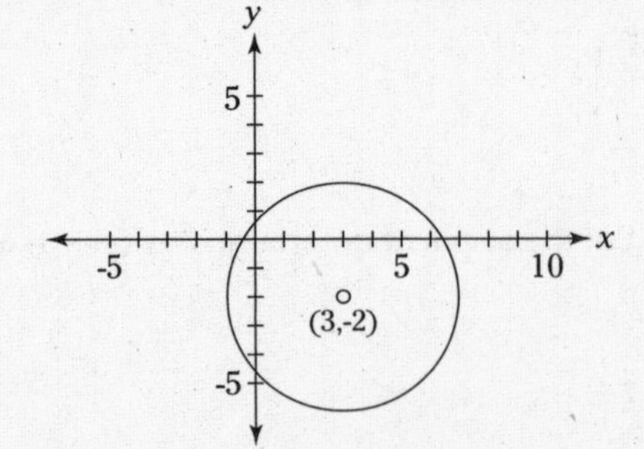

24 Sketch the graph of the circle $x^2 + (y-4)^2 = 36$. **See the following figure.**

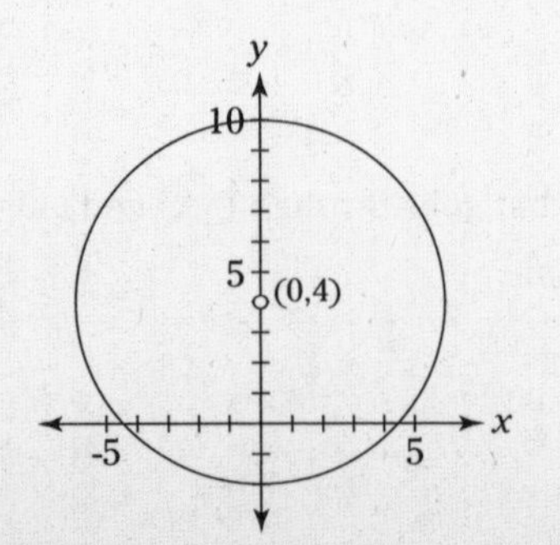

Chapter 3

Getting the Third Degree

In This Chapter

- Dealing in degrees
- Acquainting yourself with angles
- Looking at coterminal angles

Trigonometry is all about angles. Angles can be measured in degrees or radians, but the starting place is degrees, which is what this chapter is about. Angle measures are the input values for the trigonometric functions. The size of the angle determines the value of the function. An angle of 100 degrees makes some functions positive and some negative. The positions of the angles with respect to the different quadrants is what determines the signs of the different functions. This chapter acquaints you with the relative positions of the angles so that the function values will make more sense.

Recognizing First-Quadrant Angles

Before any angle measures can be covered, the *standard position* of an angle has to be established. An angle in *standard position* has its *initial* (beginning) side on the positive *x*-axis, and the *terminal* (ending) side is rotated counterclockwise from that initial side. The sweep of that terminal side determines the size of the angle formed. Figure 3-1 shows an angle of 30 degrees in standard position. Notice that the positive *x*-axis is labeled with 0 degrees, and the positive *y*-axis is labeled with 90 degrees.

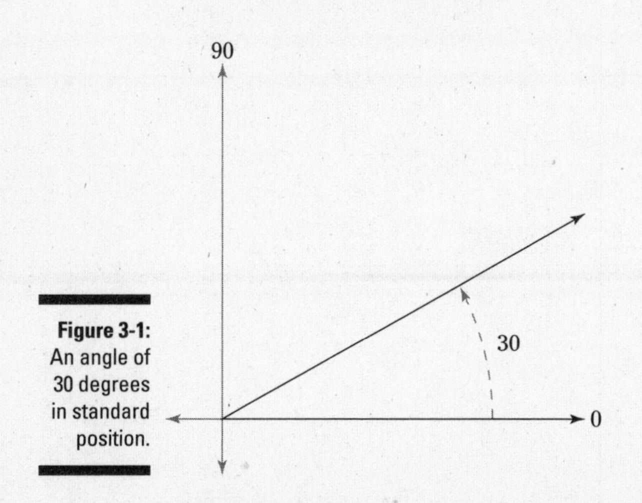

Figure 3-1: An angle of 30 degrees in standard position.

The angles in Quadrant I are greater than 0 degrees and less than 90 degrees. They're all acute angles. In Figure 3-2, rays are drawn to indicate every 15 degrees in the first quadrant. The 30-degree angle is drawn in again.

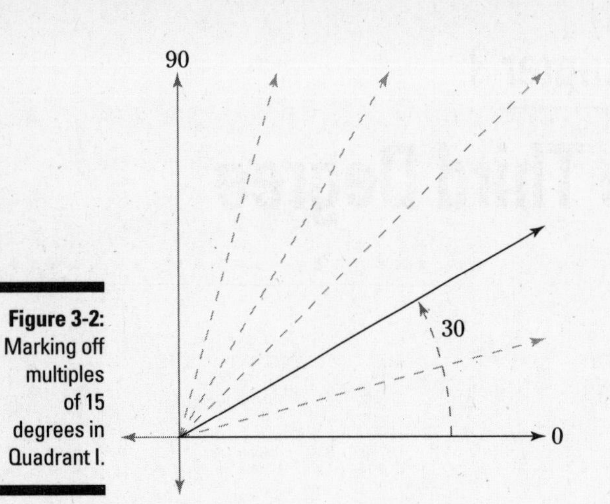

Figure 3-2: Marking off multiples of 15 degrees in Quadrant I.

EXAMPLE

Q. Is an angle, in standard position, measuring 93 degrees in Quadrant I?

A. No, this angle is too large. It has to measure less than 90 degrees.

Q. Name all the angles in Quadrant I that are multiples of 15 degrees.

A. The angles are: 15 degrees, 30 degrees, 45 degrees, 60 degrees, 75 degrees.

1. Use the graph shown in the figure, with 15-degree increments, to determine the measure of the angle in standard position.

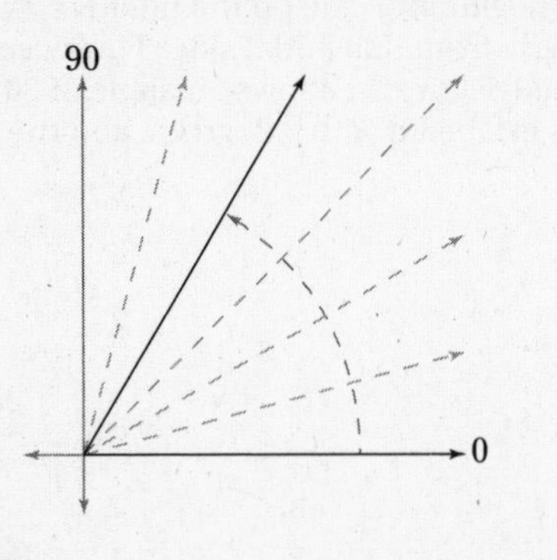

Solve It

2. Use the graph shown in the figure, with 15-degree increments, to determine the measure of the angle in standard position.

90

0

Solve It

Expanding Angles to Other Quadrants

An angle of 30 degrees is a small part of the entire picture or of a complete circle. It's a good point of reference, but there are angle measures well beyond that measure. Any angle measuring from 0 to 360 degrees can be placed in standard position before terminal sides start repeating themselves. They're all measured from the positive *x*-axis and go counterclockwise from that ray. Angles in Quadrant I are greater than 0 degrees and less than 90 degrees. Those in Quadrant II are greater than 90 degrees and less than 180 degrees. In Quadrant III, they're greater than 180 degrees and less than 270 degrees. And in Quadrant IV, they're greater than 270 degrees and less than 360 degrees. Figure 3-3 shows the entire coordinate axis system marked in 15-degree increments. I've drawn in angles measuring 135 degrees, 195 degrees, and 300 degrees.

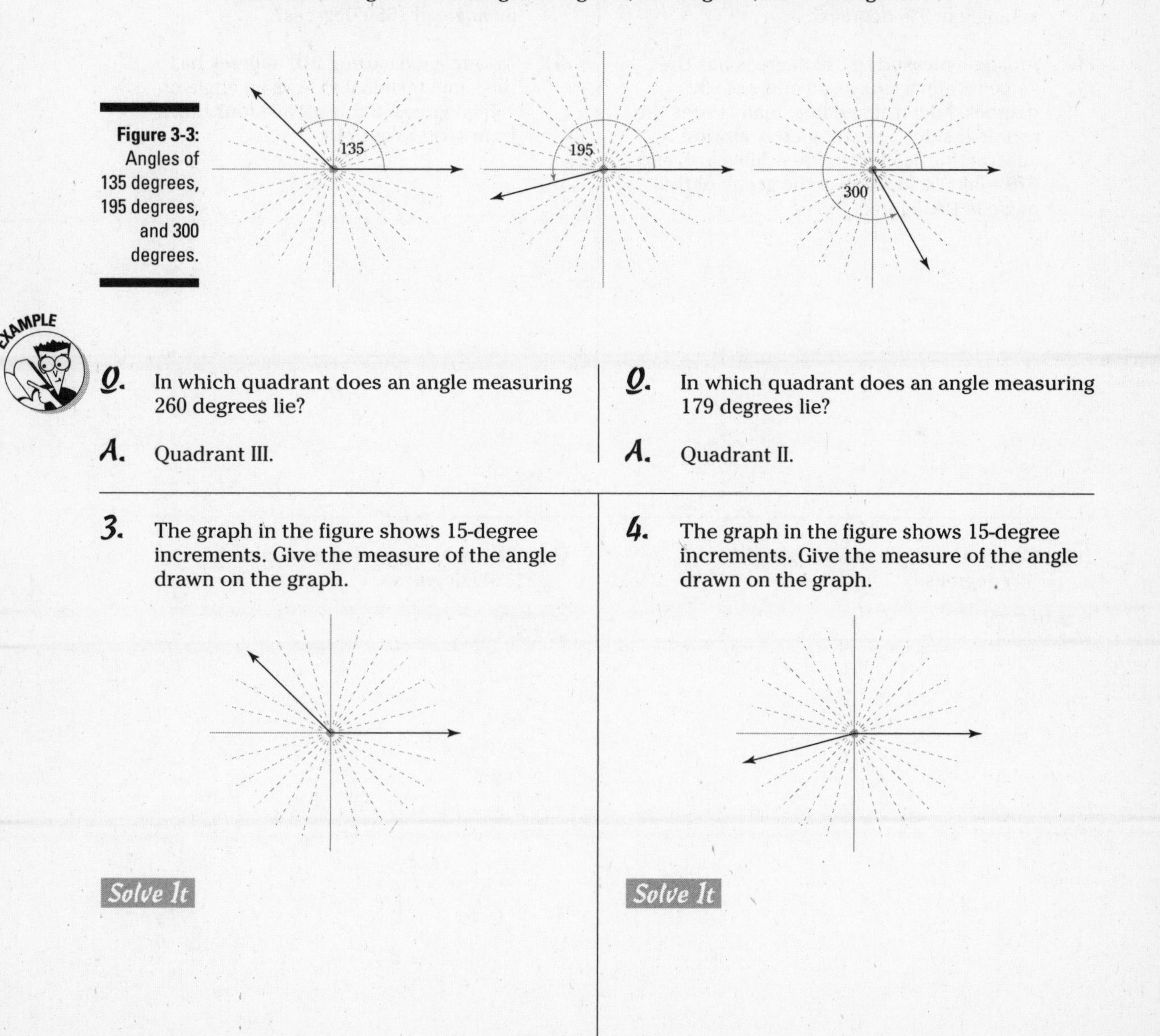

Figure 3-3: Angles of 135 degrees, 195 degrees, and 300 degrees.

Q. In which quadrant does an angle measuring 260 degrees lie?

A. Quadrant III.

Q. In which quadrant does an angle measuring 179 degrees lie?

A. Quadrant II.

3. The graph in the figure shows 15-degree increments. Give the measure of the angle drawn on the graph.

4. The graph in the figure shows 15-degree increments. Give the measure of the angle drawn on the graph.

Expanding Angles beyond 360 Degrees

Just when you thought you'd seen everything there is to see as far as angle measures, I'm here to tell you that there's more — and more and more. Angle measures aren't restricted to numbers between 0 and 360 degrees. They can go on beyond 360. It's just that angles greater than 360 degrees have to share terminal sides with other angles. I provide more information on that in the "Dealing with Coterminal Angles" section, later in this chapter.

Q. What positive angle measuring less than 360 degrees has the same terminal side as an angle of 930 degrees?

A. An angle measuring 210 degrees has the same terminal side as an angle of 930 degrees. I determined how many times the terminal side goes completely around by subtracting 360 twice. 930 – 360 = 570, and 570 – 360 = 210. Look at the graph of this angle in the figure.

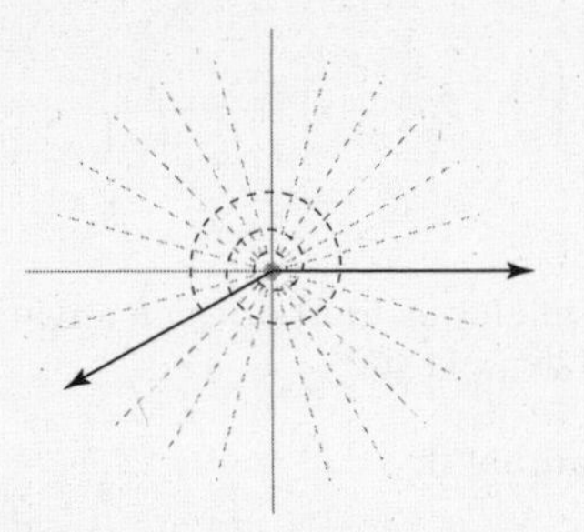

Q. What positive angle measuring less than 360 degrees has the same terminal side as an angle of 1650 degrees?

A. An angle measuring 210 degrees has the same terminal side as an angle of 1650 degrees. Subtract 360 four times from 1650 to get 210.

5. Sketch the graph of an angle of 405 degrees.

6. Sketch the graph of an angle of 1200 degrees.

Solve It

Coordinating with Negative Angle Measures

Not all angles have positive measures. An angle in standard position starts its measure on the positive *x*-axis and, if it's positive, goes counterclockwise. You can draw angles going in the other direction (clockwise) if you indicate that it's a negative measure. Why measure in the other direction? Maybe you get dizzy going counterclockwise and need to turn clockwise. Figure 3-4 shows two angles that have negative measures.

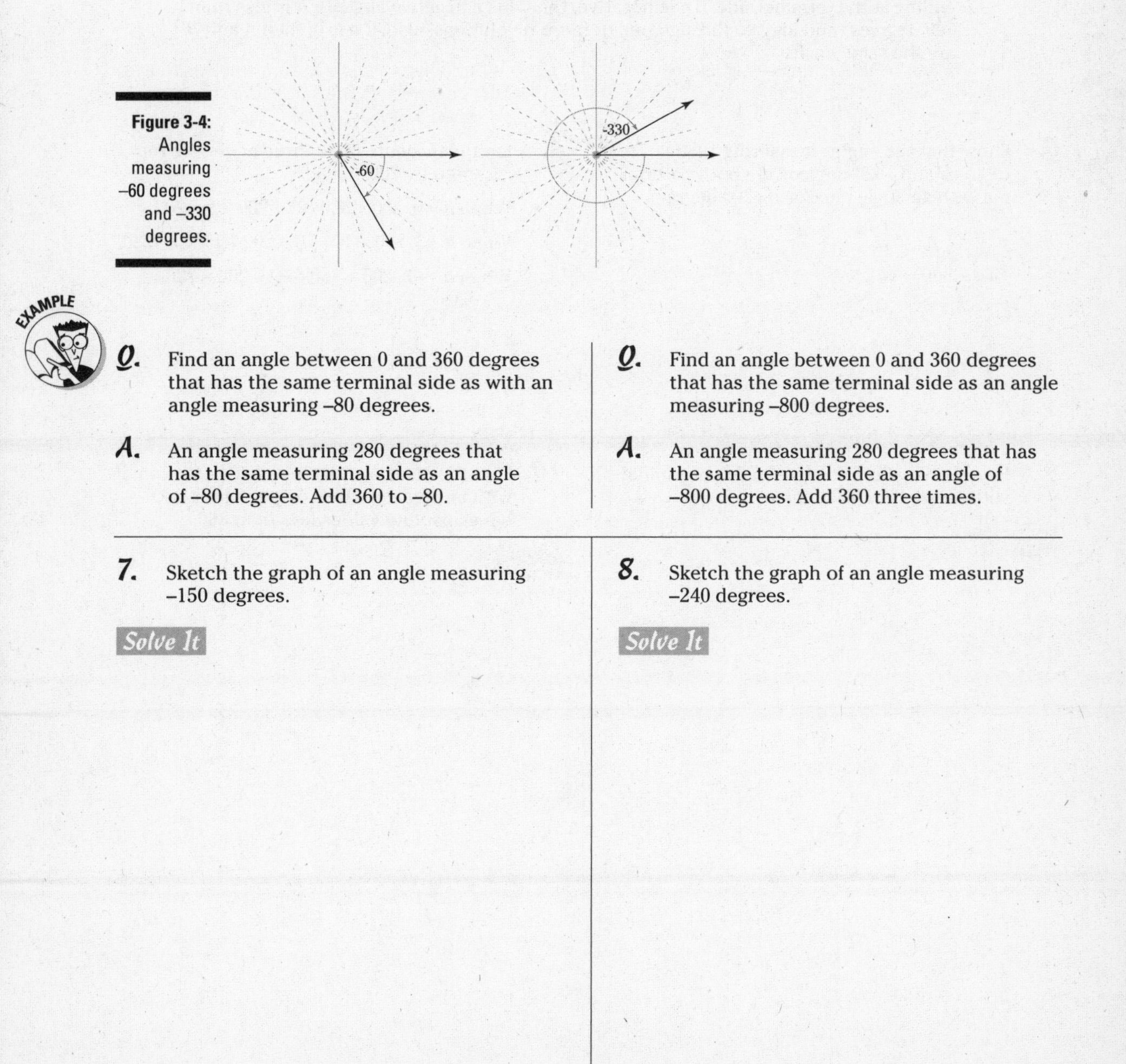

Figure 3-4: Angles measuring –60 degrees and –330 degrees.

EXAMPLE

Q. Find an angle between 0 and 360 degrees that has the same terminal side as with an angle measuring –80 degrees.

A. An angle measuring 280 degrees that has the same terminal side as an angle of –80 degrees. Add 360 to –80.

Q. Find an angle between 0 and 360 degrees that has the same terminal side as an angle measuring –800 degrees.

A. An angle measuring 280 degrees that has the same terminal side as an angle of –800 degrees. Add 360 three times.

7. Sketch the graph of an angle measuring –150 degrees.

Solve It

8. Sketch the graph of an angle measuring –240 degrees.

Solve It

Dealing with Coterminal Angles

An angle has an *initial* side and *terminal* side. When two angles share the same terminal side, they're said to be *coterminal.* Angles that are coterminal when they're in standard position have a particular relationship. Let angles A and B be coterminal. Then the measure of angle A is equal to the measure of angle $B + 360n$, where n is some integer. If n is positive, then angle A goes through one or more complete revolutions before ending at the terminal side. If n is negative, then A is a negative angle (if B is less than 360 degrees) and may go through one or more revolutions, also. If n is 0, then A and B are the same angle.

Q. Show that the angles measuring 1280 degrees and –520 degrees are both coterminal with an angle measuring 200 degrees.

A. Let the measure of all three angles be represented by $200 + 360n$:

When $n = 0$, $200 + 360(0) = 200 + 0 = 200$.

When $n = 3$, $200 + 360(3) = 200 + 1080 = 1280$.

When $n = -2$, $200 + 360(-2) = 200 - 720 = -520$

9. Find the smallest positive angle coterminal with an angle of 960 degrees.

Solve It

10. Find the measures of two angles coterminal with an angle of –1000 degrees that both have absolute values less than 360.

Solve It

Answers to Problems on Measuring in Degrees

The following are the solutions to the practice problems presented earlier in this chapter.

1. Use the graph shown in the figure, with 15-degree increments, to determine the measure of the angle in standard position. **60 degrees.**

   Each section is 15 degrees, so this angle measures 60 degrees.

2. Use the graph shown in the figure, with 15-degree increments, to determine the measure of the angle in standard position. **45 degrees.**

   Each section is 15 degrees, so this angle is 45 degrees.

3. The graph in the figure shows 15-degree increments. Give the measure of the angle drawn on the graph. **135 degrees.**

   This angle spans 9 sections, so that's 135 degrees. Another way of looking at it is that Quadrant I contains 90 degrees. This angle is 45 degrees (three sections) more than that: 90 + 45 = 135.

4. The graph in the figure shows 15-degree increments. Give the measure of the angle drawn on the graph. **195 degrees.**

   This angle spans 13 sections, so that's 195 degrees. Another way of looking at it is that Quadrants I and II contain 180 degrees. This angle is 15 degrees (one section) more than that: 180 + 15 = 195.

5. Sketch the graph of an angle of 405 degrees. **See the following figure.**

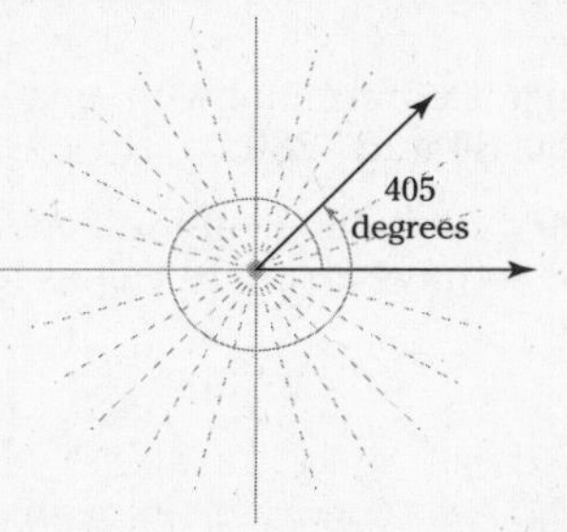

6. Sketch the graph of an angle of 1200 degrees. **See the following figure.**

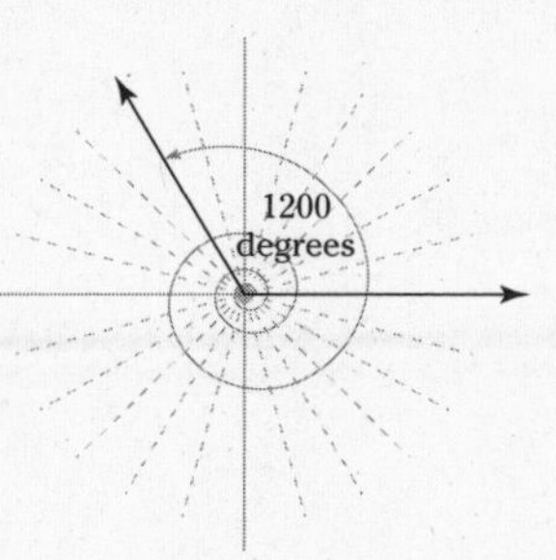

7 Sketch the graph of an angle measuring –150 degrees. **See the following figure.**

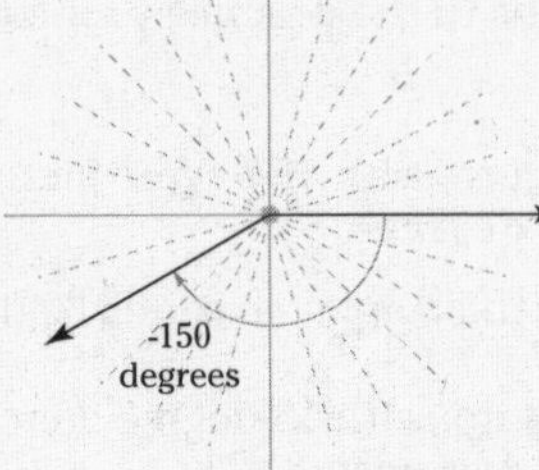

8 Sketch the graph of an angle measuring –240 degrees. **See the following figure.**

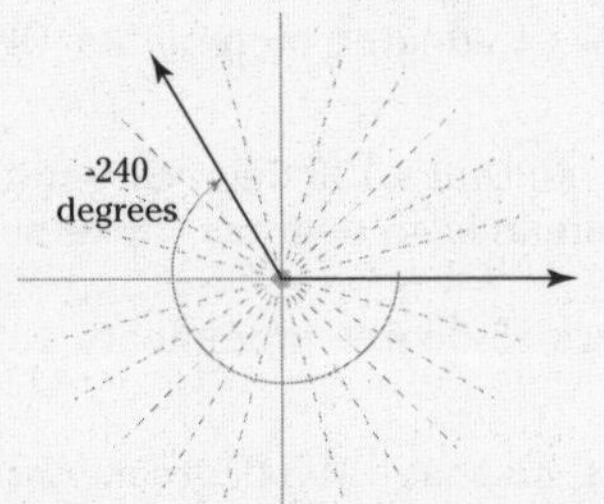

9 Find the smallest positive angle coterminal with an angle of 960 degrees. **240 degrees.**

Subtract as many multiples of 360 as possible to find the smallest positive angle: 960 – 360 (2) = 960 – 720 = 240 degrees.

10 Find the measures of two angles coterminal with an angle of –1000 degrees that both have absolute values less than 360. **80 and –280.**

First, –1000 + 360 (3) = –1000 + 1080 = 80 degrees. Second, –1000 + 360 (2) = –1000 + 720 = –280. Both of these — 80 and –280 — have absolute values less than 360.

Chapter 4

Recognizing Radian Measure

In This Chapter

- Measuring in radians
- Changing from degrees to radians and back again
- Measuring arcs and sectors

Measuring angles in degrees is how trigonometry got started. The number 360 for the number of degrees wasn't completely arbitrary, but it was man-chosen and created to be convenient. Radian measure is something more natural. It's based on the circumference of the circle and on another *natural* number, π. The advantage that radian measure has over degree measure is that radian measures are already real numbers. The degree measurement is something that doesn't combine easily with other numbers. Multiplying degrees times inches is like trying to multiply apples times cars — they just don't compute.

A radian is about 57 degrees — and *about* is the best I can do. A circle is divided into 360 degrees, and it's also divided into 2π radians — that means there are about 2(3.14) = 6.28 radians in the complete circle — as opposed to 360 smaller measures. Radian measure is really more usable in science, engineering, and other mathematical applications, because it's directly linked to the size of the circle. It's just that radians aren't as nice a number to work with.

Becoming Acquainted with Graphed Radians

Figure 4-1 shows the graph of 1 radian superimposed on degrees.

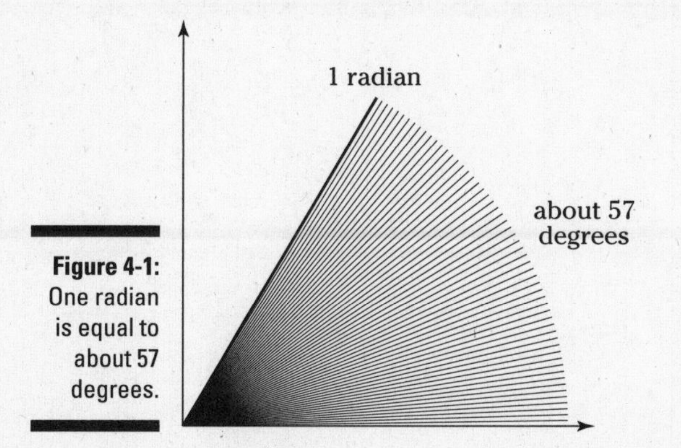

Figure 4-1: One radian is equal to about 57 degrees.

Most radian measures of angles are given in terms of a multiple of π. But it's still important to acquaint yourself with the relative measures of degrees and radians. When sketching radian measure, use an estimate of about one-sixth of the circle for each radian, and you won't go wrong.

Q. In what quadrant will an angle measuring 2 radians lie?

A. Quadrant II. Two radians is about 114 degrees.

Q. In what quadrant will an angle measuring 3.2π radians lie?

A. Quadrant III. This is slightly more than 3π radians. 3π radians is equal to 540 degrees. The terminal side of an angle of 540 degrees is the same as an angle of 180 degrees. So the angle of 3.2π goes slightly past that left axis into Quadrant III.

1. Sketch the graph of an angle measuring 3 radians.

Solve It

2. Sketch the graph of an angle measuring –2 radians.

Solve It

3. Sketch the graph of an angle measuring -2π radians.

Solve It

4. Sketch the graph of an angle measuring 8 radians.

Solve It

Changing from Degrees to Radians

Working in trigonometry, calculus, and other mathematical areas can require you to switch from degree measures to radian measures and back again. You may be lucky enough to have a calculator to do the conversion for you, but it's always a good idea to have a backup in the form of a pencil and paper for computation. A fairly easy formula will serve to change from one form to the other. You don't need to memorize two different formulas for the two different conversions.

To change from degrees to radians or vice versa, use the formula $\frac{\theta^\circ}{180} = \frac{\theta^R}{\pi}$, where θ° is the general angle theta measured in degrees, and θ^R is the same angle theta measured in radians.

Q. Change measures of 60 degrees and 330 degrees to radians.

A. First, converting the 60 degrees, replace the θ° in the formula with 60 and reduce the fraction on the left:

$$\frac{60}{180} = \frac{\theta^R}{\pi}$$

$$\frac{1}{3} = \frac{\theta^R}{\pi}$$

Then multiply each side by π to get the measure in radians:

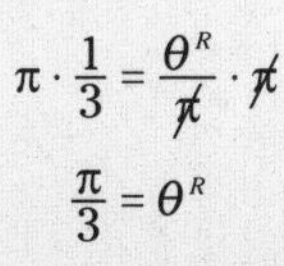

$$\pi \cdot \frac{1}{3} = \frac{\theta^R}{\not{\pi}} \cdot \not{\pi}$$

$$\frac{\pi}{3} = \theta^R$$

This angle measure is read, "Pi over three radians." Next, converting 330 degrees to radians, follow the same steps as in the previous example:

$$\frac{330}{180} = \frac{\theta^R}{\pi}$$

$$\frac{11}{6} = \frac{\theta^R}{\pi}$$

$$\pi \cdot \frac{11}{6} = \frac{\theta^R}{\not{\pi}} \cdot \not{\pi}$$

$$\frac{11\pi}{6} = \theta^R$$

5. Change 150 degrees to radians.

Solve It

6. Change 105 degrees to radians.

Solve It

7. Change 900 degrees to radians.

Solve It

8. Change –450 degrees to radians.

Solve It

Changing from Radians to Degrees

Changing from radians to degrees uses the same formula as is used to change from degrees to radians: $\frac{\theta^\circ}{180} = \frac{\theta^R}{\pi}$. The process is pretty much the same, except that when the θ^R is replaced with the angle value, often, a complex fraction occurs and has to be dealt with.

Q. Change $\frac{5\pi}{6}$ radians to degrees.

A. Substituting the radian measure into the formula, $\frac{\theta^\circ}{180} = \frac{\frac{5\pi}{6}}{\pi}$, you see the complex fraction on the right. The easiest way to deal with this is to multiply the numerator by the reciprocal of the denominator, reduce the fractions, and simplify:

$$\frac{\theta^\circ}{180} = \frac{5\not{\pi}}{6} \cdot \frac{1}{\not{\pi}} = \frac{5}{6}$$

Then multiply each side by 180 to solve for the angle measure in degrees.

$$\not{180} \cdot \frac{\theta^\circ}{\not{180}} = \frac{5}{\not{6}} \cdot \not{180}^{30}$$
$$\theta^\circ = 150$$

9. Change the angle $\frac{3\pi}{4}$ radians to degrees.

Solve It

10. Change the angle $\frac{5\pi}{3}$ radians to degrees.

Solve It

11. Change the angle $-\frac{7\pi}{6}$ radians to degrees.

Solve It

12. Change the angle 0.628 radians to degrees. (***Hint:*** The approximation of π, 3.14, has already been multiplied through to get this decimal value.)

Solve It

Measuring Arcs

The distance all the way around a circle is its *circumference.* A piece of the circumference is called an *arc.* To find the length of a particular arc, all you need is the measure of the *central angle* (an angle with its vertex at the center) that cuts off that arc, and the radius of the circle. In Figure 4-2, there's a picture of a circle with a radius of 6 inches and a central angle measuring 75 degrees. In the example, I find the length of the arc cut off by that angle. The formula for finding the length of an arc is $s = \theta r$ where θ is the central angle given in radians, and r is the radius of the circle.

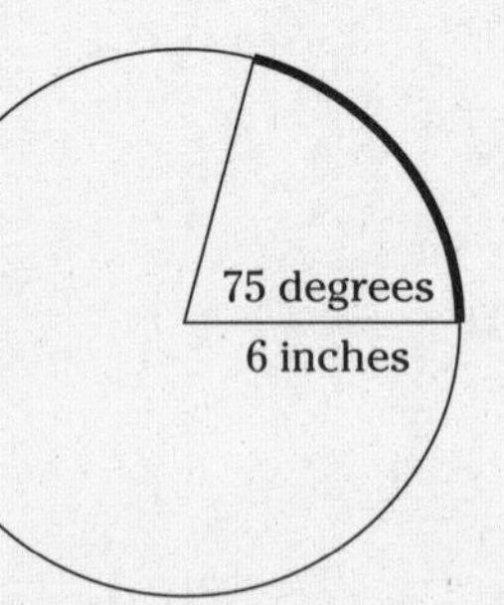

Figure 4-2: An arc in a circle with radius 6 inches, determined by a 75-degree angle.

Q. Find the length of an arc in a circle with a radius of 6 inches, if the central angle determining it measures 75 degrees.

A. To find the length of an arc in a circle with a radius of 6 inches that's determined by an angle of 75 degrees, first change the 75 degrees to radian measure:

$$\frac{75}{180} = \frac{\theta^R}{\pi}$$

$$\frac{5}{12} = \frac{\theta^R}{\pi}$$

$$\frac{5\pi}{12} = \theta^R$$

Then put that radian measure in the formula for arc length and solve. You can use 3.14 to approximate the value of π.

$$s = \frac{5\pi}{\cancel{12}_2} \cdot \cancel{6} = \frac{5\pi}{2} \approx 7.85 \text{ inches.}$$

13. Find the length of the arc in a circle with a radius of 2 feet, if the central angle determining it measures 60 degrees.

Solve It

14. Find the length of the arc in a circle with a diameter of 8 feet, if the central angle determining it measures 330 degrees.

Solve It

Determining the Area of a Sector

A sector of a circle is like a slice of pie or a wedge of a round cheese. In Chapter 1, you'll find a method for finding the area of a sector of a circle by figuring out what fraction of the circle you're dealing with, and then multiplying by the area of the entire circle. Radians make the job easier, because they fit right into a formula for the area of a sector. You can't multiply by degrees — the radians are real numbers and combine with other real numbers. The formula is: $A = \frac{1}{2}\theta r^2$ where θ is in radians.

Q. Which would you rather have, a piece of an 8-inch pie that's been cut into sixths or a piece of a 10-inch pie that's been cut into eighths?

A. This is a problem involving sectors (see the figure). One-sixth of a pie is $\frac{1}{6}$ of 2π radians. The measure of the central angle is $\frac{1}{6}\cdot 2\pi = \frac{2\pi}{6} = \frac{\pi}{3}$. An 8-inch pie has a 4-inch radius. Putting the angle and radius into the formula for the area of a sector, $A = \frac{1}{2}\cdot\frac{\pi}{3}\cdot 4^2 = \frac{16\pi}{6} = \frac{8\pi}{3} \approx 8.38$ square inches. One-eighth of a pie is $\frac{1}{8}$ of 2π radians. The measure of the angle is $\frac{1}{8}\cdot 2\pi = \frac{2\pi}{8} = \frac{\pi}{4}$. A 10-inch pie has a 10-inch pie has a 5-inch radius. Putting the angle and radius into the formula for the area of a sector, $A = \frac{1}{2}\cdot\frac{\pi}{4}\cdot 5^2 = \frac{25\pi}{8} \approx 9.82$ square inches. There isn't too much difference, but it looks like the smaller part of the bigger pie has the larger piece, in terms of area.

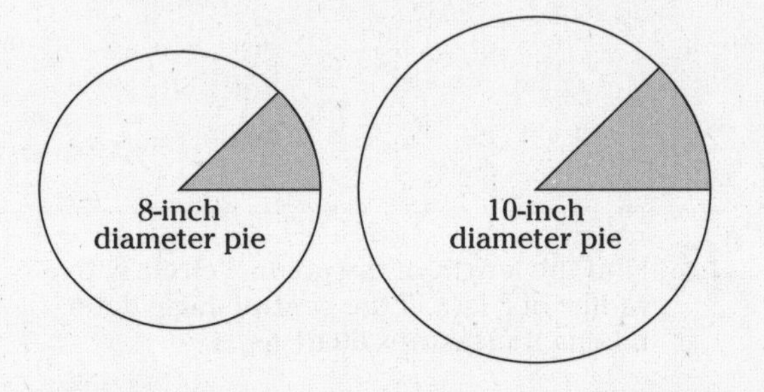

15. Find the area of the sector of a circle with a 9-inch radius if the central angle of the sector is $\frac{5\pi}{6}$.

16. Find the area of the sector of a circle with a 12-inch diameter if the central angle of the sector measures 120 degrees.

Answers to Problems on Radian Measure

The following are the solutions to the practice problems presented earlier in this chapter.

1. Sketch the graph of an angle measuring 3 radians. **See the following figure.**

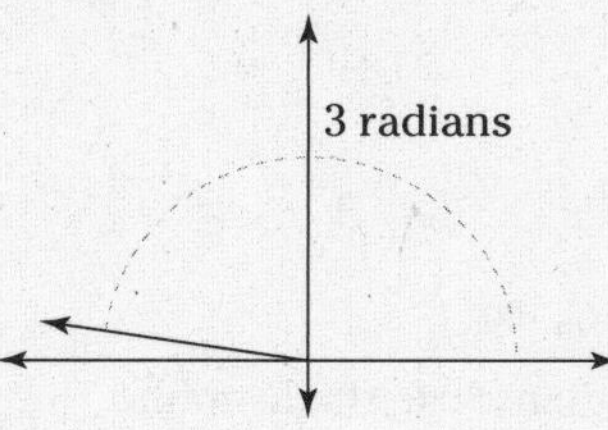

2. Sketch the graph of an angle measuring –2 radians. **See the following figure.**

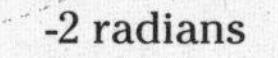

3. Sketch the graph of an angle measuring -2π radians. **See the following figure.**

   This is a complete rotation, in clockwise direction.

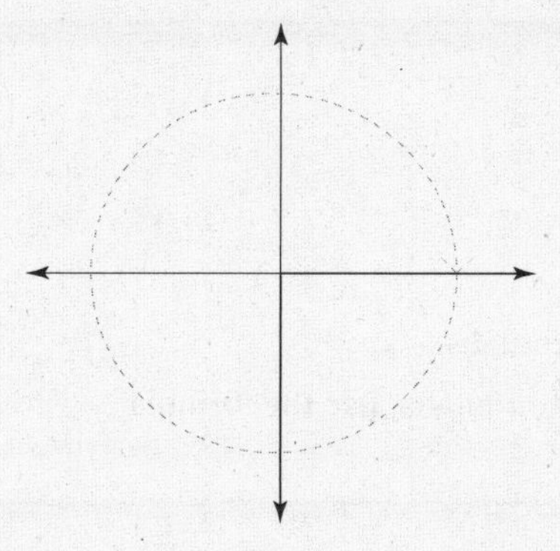

4. Sketch the graph of an angle measuring 8 radians. **See the following figure.**

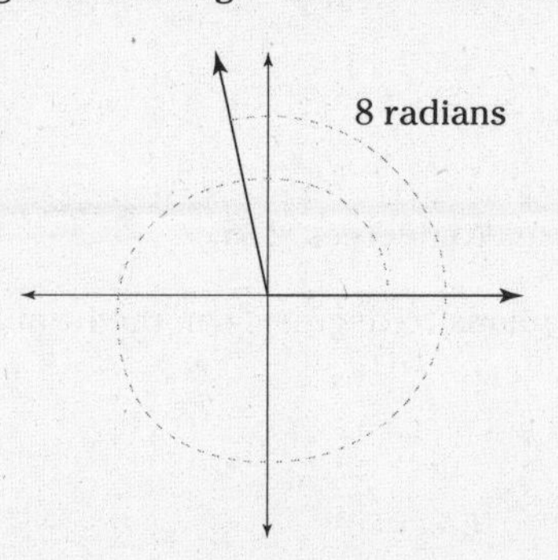

5. Change 150 degrees to radians. **$\frac{5\pi}{6}$.**

To change 150 degrees to radians, use the formula:

$$\frac{150}{180} = \frac{\theta^R}{\pi}$$
$$\frac{5}{6} = \frac{\theta^R}{\pi}$$
$$\pi \cdot \frac{5}{6} = \frac{\theta^R}{\cancel{\pi}} \cdot \cancel{\pi}$$
$$\frac{5\pi}{6} = \theta^R$$

6. Change 105 degrees to radians. **$\frac{7\pi}{12}$.**

To change 105 degrees to radians, use the formula:

$$\frac{105}{180} = \frac{\theta^R}{\pi}$$
$$\frac{7}{12} = \frac{\theta^R}{\pi}$$
$$\pi \cdot \frac{7}{12} = \frac{\theta^R}{\cancel{\pi}} \cdot \cancel{\pi}$$
$$\frac{7\pi}{12} = \theta^R$$

7. Change 900 degrees to radians. **5π.**

To change 900 degrees to radians, use the formula:

$$\frac{900}{180} = \frac{\theta^R}{\pi}$$
$$5 = \frac{\theta^R}{\pi}$$
$$\pi \cdot 5 = \frac{\theta^R}{\cancel{\pi}} \cdot \cancel{\pi}$$
$$5\pi = \theta^R$$

8. Change –450 degrees to radians. **$-\frac{5\pi}{2}$.**

To change –450 degrees to radians, use the formula:

$$\frac{-450}{180} = \frac{\theta^R}{\pi}$$
$$-\frac{5}{2} = \frac{\theta^R}{\pi}$$
$$\pi\left(-\frac{5}{2}\right) = \frac{\theta^R}{\cancel{\pi}} \cdot \cancel{\pi}$$
$$-\frac{5\pi}{2} = \theta^R$$

9. Change the angle $\frac{3\pi}{4}$ radians to degrees. **135.**

To change the angle $\frac{3\pi}{4}$ radians to degrees, use the formula:

$$\frac{\theta^\circ}{180} = \frac{\frac{3\pi}{4}}{\pi}$$
$$\frac{\theta^\circ}{180} = \frac{3\pi}{4} \cdot \frac{1}{\pi}$$
$$\frac{\theta^\circ}{180} = \frac{3}{4}$$
$$180 \cdot \frac{\theta^\circ}{180} = \frac{3}{4} \cdot 180$$
$$\theta^\circ = 135$$

10 Change the angle $\frac{5\pi}{3}$ radians to degrees. **300.**

To change the angle $\frac{5\pi}{3}$ radians to degrees, use the formula:

$$\frac{\theta^\circ}{180}=\frac{\frac{5\pi}{3}}{\pi}$$
$$\frac{\theta^\circ}{180}=\frac{5\pi}{3}\cdot\frac{1}{\pi}$$
$$\frac{\theta^\circ}{180}=\frac{5}{3}$$
$$180\cdot\frac{\theta^\circ}{180}=\frac{5}{3}\cdot 180$$
$$\theta^\circ=300$$

11 Change the angle $-\frac{7\pi}{6}$ radians to degrees. **–210.**

To change the angle $-\frac{7\pi}{6}$ radians to degrees, use the formula:

$$\frac{\theta^\circ}{180}=\frac{-\frac{7\pi}{6}}{\pi}$$
$$\frac{\theta^\circ}{180}=-\frac{7\pi}{6}\cdot\frac{1}{\pi}$$
$$\frac{\theta^\circ}{180}=-\frac{7}{6}$$
$$180\cdot\frac{\theta^\circ}{180}=-\frac{7}{6}\cdot 180$$
$$\theta^\circ=-210$$

12 Change the angle 0.628 radians to degrees. (***Hint:*** The approximation of π, 3.14, has already been multiplied through to get this decimal value.) **36.**

To change the angle 0.628 radians to degrees, use the same formula as for the other angles: $\frac{\theta^\circ}{180}=\frac{0.628}{\pi}$. Now replace the π with 3.14 and simplify:

$$\frac{\theta^\circ}{180}=\frac{0.628}{3.14}$$
$$\frac{\theta^\circ}{180}=0.2$$
$$180\cdot\frac{\theta^\circ}{180}=0.2\,(180)$$
$$\theta^\circ=36$$

13 Find the length of the arc in a circle with a radius of 2 feet, if the central angle determining it measures 60 degrees. ≈ **2.09 feet.**

To find the length of the arc in a circle with a radius of 2 feet, if the central angle determining it measures 60 degrees, you need to use the formula for arc length. In that case, the angle has to be in radians. Change 60 degrees to radians:

$$\frac{60}{180}=\frac{\theta^R}{\pi}$$
$$\frac{1}{3}=\frac{\theta^R}{\pi}$$
$$\pi\cdot\frac{1}{3}=\frac{\theta^R}{\pi}\cdot\pi$$
$$\frac{\pi}{3}=\theta^R$$

Now, use the formula for arc length, $s=\frac{\pi}{3}\cdot 2=\frac{2\pi}{3}\approx 2.09$ feet.

14 Find the length of the arc in a circle with a diameter of 8 feet, if the central angle determining it measures 330 degrees. ≈ **23.04 feet.**

To find the length of the arc in a circle with a diameter of 8 feet, if the central angle determining it measures 330 degrees, first you need to change the 330 degrees to radians:

$$\frac{330}{180} = \frac{\theta^R}{\pi}$$
$$\frac{11}{6} = \frac{\theta^R}{\pi}$$
$$\pi \cdot \frac{11}{6} = \frac{\theta^R}{\pi} \cdot \pi$$
$$\frac{11\pi}{6} = \theta^R$$

You need the radius, not the diameter. Half of 8 is 4, the radius. Now, use that in the formula for the arc length: $s = \frac{11\pi}{6} \cdot 4 = \frac{22\pi}{3} \approx 23.04$.

15 Find the area of the sector of a circle with a 9-inch radius if the central angle of the sector is $\frac{5\pi}{6}$. ≈ **106.03 square inches.**

Use the formula for the area of a sector, $A = \frac{1}{2}\left(\frac{5\pi}{6}\right)9^2 = \frac{5\pi}{12} \cdot 81 = \frac{135\pi}{4} \approx 106.03$ square inches.

I used $\pi \approx 3.1416$ to get the answer. You will also get this answer using the built in π value in your calculator.

16 Find the area of the sector of a circle with a 12-inch diameter if the central angle of the sector measures 120 degrees. ≈ **37.70 square inches.**

To find the area of the sector of a circle with a 12-inch diameter if the central angle of the sector measures 120 degrees, first change the 120 degrees to radians:

$$\frac{120}{180} = \frac{\theta^R}{\pi}$$
$$\frac{2}{3} = \frac{\theta^R}{\pi}$$
$$\pi \cdot \frac{2}{3} = \frac{\theta^R}{\pi} \cdot \pi$$
$$\frac{2\pi}{3} = \theta^R$$

Use this radian measure and the radius of the circle, 6 inches, in the formula for the area of a sector: $A = \frac{1}{2}\left(\frac{2\pi}{3}\right)6^2 = \frac{\pi}{3} \cdot 36 = 12\pi \approx 37.70$ square inches.

I used $\pi \approx 3.1416$ to get the answer. You will also get this answer using the built in π value in your calculator.

Chapter 5

Making Things Right with Right Triangles

In This Chapter

- Examining the right triangle
- Using Pythagorean triples
- Identifying special right triangles
- Applying the right triangle to practical applications

The right triangle can take on many different sizes and shapes, but all right triangles have one important characteristic: They have a right angle — the two shorter sides of the triangle meet to form a 90-degree angle. Triangles can have only one right angle. The other two angles are acute angles. The two acute angles are also complementary, and some interesting properties arise in the trig functions because of this.

Naming the Parts of a Right Triangle

A right triangle has six parts, like any other triangle. Special designations for right triangles are used to identify particular properties. The two shorter sides — the ones opposite the acute angles — are called the *legs.* The longest side is called the *hypotenuse.* The hypotenuse is always opposite the right angle. The legs also have special names, when they're being considered with respect to a particular acute angle. A leg is called *opposite* or *adjacent,* depending on whether it's not a part of the acute angle *(opposite)* or whether it's one of the sides of the acute angle *(adjacent).* And the names change for the legs, depending on which acute angle you're talking about at the time.

Look at the triangle *ABC* in Figure 5-1 and the different descriptions that can arise. The segment *CD* is perpendicular to the side *AB,* creating two right triangles within a right triangle.

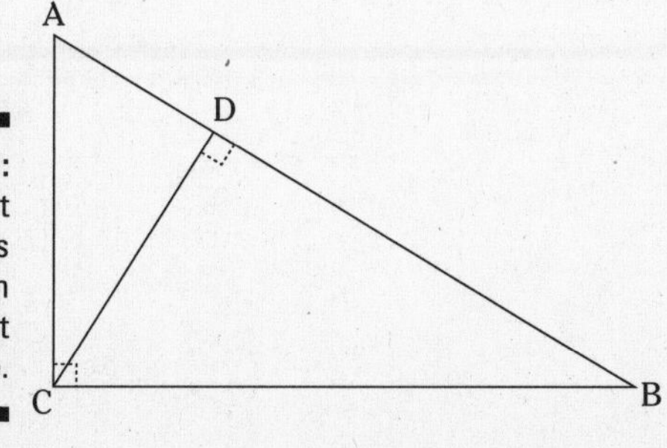

Figure 5-1: Two right triangles within a right triangle.

In Figure 5-1, when considering acute angle B in triangle ABC, side AC is *opposite* and side BC is *adjacent* to angle B. When considering acute angle A in triangle ABC, side BC is *opposite* and side AC is *adjacent* to angle A.

In triangle ABC, side AB is the *hypotenuse*. In triangle ACD, side AC is the *hypotenuse*. In triangle BCD, side BC is the *hypotenuse*.

Q. In Figure 5-1, look at triangle ACD. What side is *opposite* angle ACD in that triangle?

A. Side AD is opposite angle ACD.

Q. In Figure 5-1, look at triangle BCD. What side is *opposite* angle B in that triangle?

A. Side CD is opposite angle B.

1. In Figure 5-1, what side is *opposite* angle A in right triangle ACD?

Solve It

2. In Figure 5-1, what side is *adjacent* to angle B, in right triangle BCD?

Solve It

3. In Figure 5-1, name the three right angles.

Solve It

4. In Figure 5-1, what acute angle in triangle *ACD* is the same measure as angle *BCD* in triangle *BCD?* (The three triangles are all similar to one another — they have angles that correspond and are equal to one another.)

Completing Pythagorean Triples

A Pythagorean triple is a listing of three numbers that satisfy the Pythagorean theorem. Some examples of Pythagorean triples are: {3, 4, 5}, {5, 12, 13}, and {21, 220, 221}. A Pythagorean triple provides a solution to $a^2 + b^2 = c^2$. There's a great formula that will generate an infinite number of Pythagorean triples that are integers (not fractions). You can choose any two positive integers — call them *s* and *t*. Let *s* be larger than *t*. Then the following three numbers are a Pythagorean triple: $\{2st,\ s^2 - t^2,\ s^2 + t^2\}$. For instance, if $s = 6$ and $t = 1$, then $2st = 12$, $s^2 - t^2 = 36 - 1 = 35$ and $s^2 + t^2 = 36 + 1 = 37$. You get the Pythagorean triple: {12, 35, 37}. Substituting those numbers into the Pythagorean theorem:

$$12^2 + 35^2 = 37^2$$
$$144 + 1225 = 1369$$

Q. Find any Pythagorean triples that occur when $st = 12$.

A. The product $st = 12$ occurs when $s = 12$, $t = 1$; or $s = 6$, $t = 2$; or $s = 4$, $t = 3$. So there are three Pythagorean triples to create.

When $s = 12$, $t = 1$, then $2st = 24$, $s^2 - t^2 = 144 - 1 = 143$, and $s^2 + t^2 = 144 + 1 = 145$. The triple: {24, 143, 145}.

When $s = 6$, $t = 2$, then $2st = 24$, $s^2 - t^2 = 36 - 4 = 32$, and $s^2 + t^2 = 36 + 4 = 40$. The triple: {24, 32, 40}.

When $s = 4$, $t = 3$, then $2st = 24$, $s^2 - t^2 = 16 - 9 = 7$, and $s^2 + t^2 = 16 + 9 = 25$. The triple: {24, 7, 25}.

5. Use the formula for finding a Pythagorean triple when $s = 5$ and $t = 3$.

Solve It

6. Use the formula for finding a Pythagorean triple when $s = 9$ and $t = 2$.

Solve It

7. Use the formula to find all Pythagorean triples that occur when $st = 6$.

Solve It

8. Use the formula to find all Pythagorean triples that occur when $st = 70$.

Solve It

Completing Right Triangles

The three sides of a right triangle must satisfy the Pythagorean theorem. This is very useful to carpenters who can use this to be sure that the sides of a house meet in a right angle. They'll measure the distance from the corner to a designated point on one wall, then the distance from the same corner to a designated point on the other wall (all of this at the same level above the ground). Then, measuring from point to point, they expect to get a value that will complete the Pythagorean theorem or they *square the corner.*

Q. A carpenter has made the measurements in the figure. Check to see if the corner is square — do the numbers satisfy the Pythagorean theorem?

14 feet, 7 inches

4 feet, 1 inch

14 feet

A. Inserting the numbers into the Pythagorean theorem:

$$14^2+\left(4\frac{1}{12}\right)^2 \stackrel{?}{=} \left(14\frac{7}{12}\right)^2$$

$$14^2+\left(\frac{49}{12}\right)^2 \stackrel{?}{=} \left(\frac{175}{12}\right)^2$$

$$196+\frac{2401}{144} \stackrel{?}{=} \frac{30625}{144}$$

$$\frac{28224}{144}+\frac{2401}{144} \stackrel{?}{=} \frac{30625}{144}$$

$$\frac{30625}{144}=\frac{30625}{144}$$

This does work. The corner is a right angle. Of course, a clever carpenter would pick measures for the two legs that are known values in a Pythagorean triple, such as 3 feet and 4 feet — and would expect a hypotenuse of 5 feet.

9. Find the missing measure in the right triangle shown in the figure.

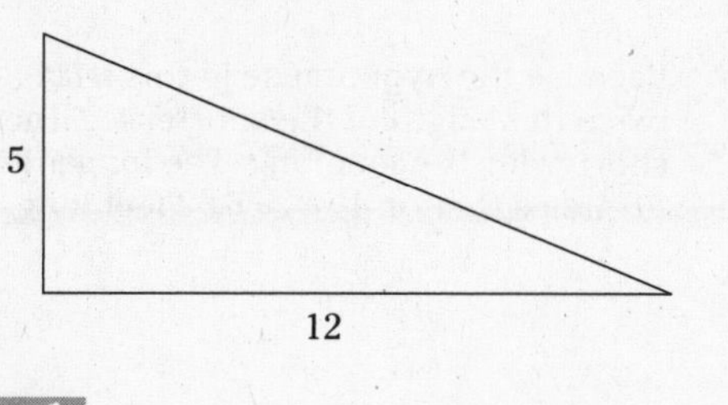

Solve It

10. Find some measures that complete the right triangle shown in the figure, using the formula for the value of the hypotenuse and determining a pair of perfect squares that work.

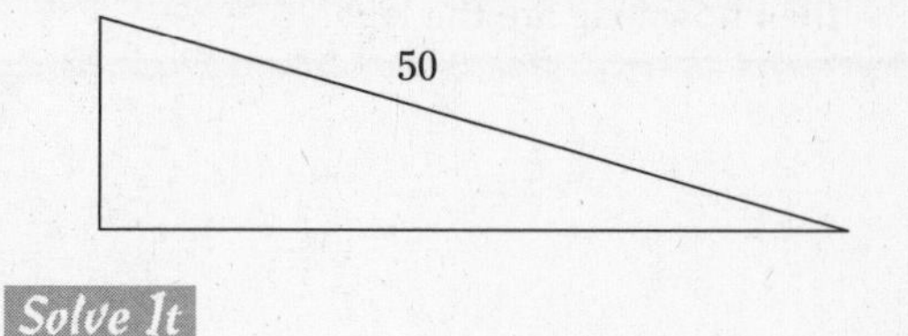

Solve It

Working with the 30-60-90 Right Triangle

One of the two special types of right triangles is the 30-60-90 right triangle. What makes it special is the relationship between the lengths of the sides. Oh, there's always the relationship of the Pythagorean theorem, but something even more special occurs in these triangles.

In a 30-60-90 right triangle, the hypotenuse is always twice the length of the shorter leg (the one opposite the 30-degree angle), and the longer leg (opposite the 60-degree angle) is always $\sqrt{3}$ times as long as the shorter leg (that's about 1.73 times as long).

Figure 5-2 shows a 30-60-90 right triangle and the relative lengths of the sides.

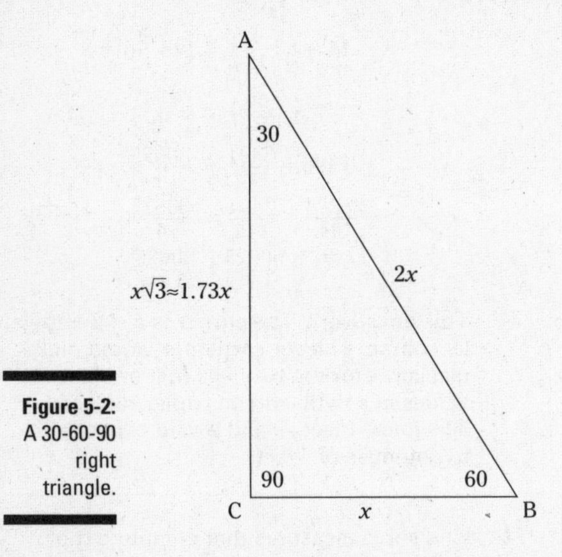

Figure 5-2: A 30-60-90 right triangle.

Q. If you know that the hypotenuse of a 30-60-90 right triangle measures 8 feet, then how long are the legs?

A. Because the hypotenuse in this triangle is twice the length of the shorter leg, then the shorter leg is 4 feet long. The longer leg is $\sqrt{3}$ times that or $4\sqrt{3} \approx 6.92$ feet long.

11. Find the measures of the other two sides of a 30-60-90 right triangle if the shorter leg is 2 inches long.

Solve It

12. Find the measures of the other two sides of a 30-60-90 right triangle if the longer leg is $7\sqrt{3}$ feet long.

13. Find the measures of the other two sides of a 30-60-90 right triangle if the shorter leg is $3\sqrt{3}$ yards long.

14. Find the measures of the other two sides of a 30-60-90 right triangle if the longer leg is 6 inches long.

Solve It

Using the Isosceles Right Triangle

The other special right triangle is the isosceles right triangle, otherwise known as the 45-45-90 right triangle. There are special relationships between the sides of this triangle — in addition to the Pythagorean theorem relationship.

In an isosceles right triangle, the length of the hypotenuse is always $\sqrt{2}$ times the length of either leg (which, of course, are the same measure, because the triangle is isosceles). This means that the hypotenuse is about 1.41 times as large as the legs.

Figure 5-3 illustrates the 45-45-90 right triangle.

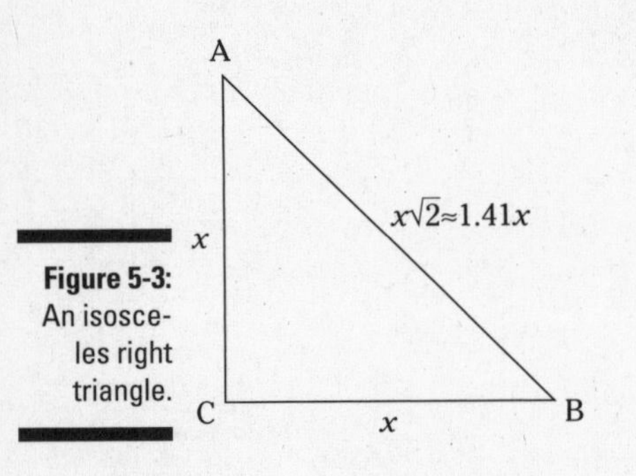

Figure 5-3: An isosceles right triangle.

Q. You're told that the perimeter of an isosceles right triangle is $2 + 2\sqrt{2}$ inches. How long are the sides of the triangle?

A. You may be tempted to say that the two legs are each 2 inches, because there's the $2\sqrt{2}$ in the sum. But if that were the case, then the sum of the two legs would be 4, and the perimeter would be $2 + 2 + 2\sqrt{2} = 4 + 2\sqrt{2}$. The other way to get the $2\sqrt{2}$ in the perimeter is if each of the legs is $\sqrt{2}$ inches long. Then you'd have, for the perimeter, $\sqrt{2} + \sqrt{2} + \sqrt{2}\sqrt{2} = 2\sqrt{2} + 2$. This seems to be it.

15. Find the measures of the two legs of the isosceles right triangle that has a hypotenuse measuring 10 feet.

Solve It

16. Find the measures of the three sides of the isosceles right triangle that has an area of 18 square yards.

Solve It

Using Right Triangles in Applications

Now it's time to make use of these right triangles and show how they can solve practical problems. The right triangle can be used if you know the measures of two sides and need a third. Also, if you have one of the special right triangles (the angle measures are available) you just need the measure of one of the sides to get the others.

Q. Clay has sighted his cat at the top of a tree. He needs to find a ladder long enough to retrieve the darling kitty. The tree is 15 feet tall, and Clay will put the foot of the ladder in a secure spot that's 8 feet from the base of the tree. How long a ladder does he need to reach the top of the tree? Look at the figure, which shows the relative measures.

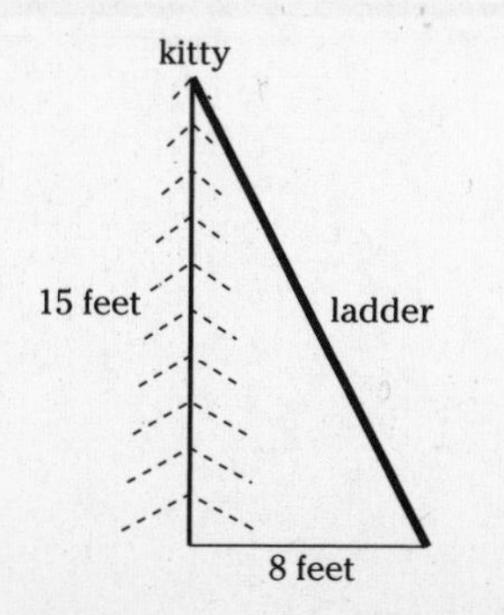

A. The measures 8 and 15 are the legs of a right triangle. Using the Pythagorean theorem:

$$8^2 + 15^2 = c^2$$
$$64 + 225 = c^2$$
$$289 = c^2$$
$$c = \sqrt{289} = 17$$

So Clay needs a ladder that's 17 feet long. Of course, by the time he finds one and puts it into place, the kitty will have climbed down by herself. They tend to do that.

17. Hank is flying a kite and has his entire 100 feet of string out. Hershel is directly under Hank's kite and is standing 28 feet from him. How high is the kite?

18. How long a fishing rod can you put into a box that's 3 feet by 5 feet by 8 feet? You need to find the diagonal length from one corner to its opposite.

19. Gladys is at an art museum and getting a sore neck from looking up and down at the pictures. When she stands in front of a picture that's 6 feet high (its length is 6 feet), she can see 30 degrees above her eye level without tilting her head and she can see 60 degrees below eye level without tilting her head. How far from the picture must she stand to be able to see the whole thing and not move her head?

20. What is the area of a regular *hexagon* (a six-sided polygon with equal sides) when its sides are 4 feet long? (***Hint:*** Subdivide it into six equilateral triangles.)

21. Stephanie walks 3 miles east and then 4 miles north. Chelsea joins her, but she walks on the diagonal (the hypotenuse of the right triangle). Chelsea sits and waits while Stephanie walks another 12 miles east and 5 miles north. Again, Chelsea joins her by walking the diagonal. How much farther did Stephanie walk than Chelsea?

Solve It

22. Crystal sights the top of a 100-foot tower and notes that the angle formed from the ground to the top of the tower is 45 degrees. David is on the opposite side of the tower from Crystal, sights the top of the tower, and notes that the angle formed from the ground to the top of the tower is 30 degrees. How far apart are Crystal and David?

Solve It

Answers to Problems on Right Triangles

The following are the solutions to the practice problems presented earlier in this chapter.

1. In Figure 5-1, what side is *opposite* angle A in right triangle ACD? **Side CD.**

2. In Figure 5-1, what side is *adjacent* to angle B, in right triangle BCD? **Side BD.**

   The side that is *adjacent* to angle B in right triangle BCD is side BD. Side BC is the hypotenuse, because it's opposite the right angle.

3. In Figure 5-1, name the three right angles. **Angle ACB, angle ADC, and angle BDC.**

   The vertex of the angle is always named as the middle letter.

4. In Figure 5-1, what acute angle in triangle ACD is the same measure as angle BCD in triangle BCD? (The three triangles are all similar to one another — they have angles that correspond and are equal to one another.) **Angle A.**

   Both of the angles, BCD in triangle BCD and A in triangle ACD, are between the hypotenuse and the shorter leg in their respective triangles.

5. Use the formula for finding a Pythagorean triple when $s = 5$ and $t = 3$. **{30, 16, 34}.**

   $2st = 2(5)(3) = 30$

   $s^2 - t^2 = 5^2 - 3^2 = 25 - 9 = 16$

   $s^2 + t^2 = 25 + 9 = 34$

   The triple is {30, 16, 34}.

6. Use the formula for finding a Pythagorean triple when $s = 9$ and $t = 2$. **{36, 77, 85}.**

   $2st = 2(9)(2) = 36$

   $s^2 - t^2 = 9^2 - 2^2 = 81 - 4 = 77$

   $s^2 + t^2 = 81 + 4 = 85$

   The triple is {36, 77, 85}.

7. Use the formula to find all Pythagorean triples that occur when $st = 6$. **{12, 35, 37} and {12, 5, 13}.**

   There are two possibilities for the values of s and t: $s = 6$, $t = 1$ or $s = 3$, $t = 2$:

   $s = 6$, $t = 1$ means that

   $2st = 2(6)(1) = 12$

   $s^2 - t^2 = 6^2 - 1^2 = 36 - 1 = 35$

   $s^2 + t^2 = 36 + 1 = 37$

   The triple is {12, 35, 37}.

   $s = 3$, $t = 2$, means that

   $2st = 2(3)(2) = 12$

   $s^2 - t^2 = 3^2 - 2^2 = 9 - 4 = 5$

   $s^2 + t^2 = 9 + 4 = 13$

   The triple is {12, 5, 13}.

8 Use the formula to find all Pythagorean triples that occur when $st = 70$. **{140, 4899, 4901}, {140, 1221, 1229}, {140, 171, 221}, and {140, 51, 149}.**

There are four possibilities for the values of *s* and *t*: $s = 70$, $t = 1$; or $s = 35$, $t = 2$; or $s = 14$, $t = 5$; or $s = 10$, $t = 7$.

$s = 70$, $t = 1$ means that

$$2st = 2(70)(1) = 140$$
$$s^2 - t^2 = 70^2 - 1^2 = 4900 - 1 = 4899$$
$$s^2 + t^2 = 4900 + 1 = 4901$$

The triple is {140, 4899, 4901}.

$s = 35$, $t = 2$, means that

$$2st = 2(35)(2) = 140$$
$$s^2 - t^2 = 35^2 - 2^2 = 1225 - 4 = 1221$$
$$s^2 + t^2 = 1225 + 4 = 1229$$

The triple is {140, 1221, 1229}.

$s = 14$, $t = 5$ means that

$$2st = 2(14)(5) = 140$$
$$s^2 - t^2 = 14^2 - 5^2 = 196 - 25 = 171$$
$$s^2 + t^2 = 196 + 25 = 221$$

The triple is {140, 171, 221}.

$s = 10$, $t = 7$, means that

$$2st = 2(10)(7) = 140$$
$$s^2 - t^2 = 10^2 - 7^2 = 100 - 49 = 51$$
$$s^2 + t^2 = 100 + 49 = 149$$

The triple is {140, 51, 149}.

9 Find the missing measure in the right triangle shown in the figure. **13.**

Use the Pythagorean theorem:

$$5^2 + 12^2 = c^2$$
$$25 + 144 = c^2$$
$$169 = c^2$$
$$13 = c$$

10 Find some measures that complete the right triangle shown in the figure, using the formula for the value of the hypotenuse and determining a pair of perfect squares that work. **14 and 48.**

You need to find *s* and *t* such that $s^2 + t^2 = 50$. The perfect squares less than 50 are: 1, 4, 9, 16, 25, 36, 49. The two that add up to 50 are 49 and 1. This is the only answer that has both *s* and *t* as integers. Using the 49 and 1, $s = 7$ and $t = 1$. $2st = 14$, and $s^2 - t^2 = 49 - 1 = 48$. So the three sides are: 50, 14, and 48. As it turns out, there are rational numbers that work. If you square 3.4 and 6.2, their sum is 50. The sides of this triangle are: 50, 42.16, and 26.88. There are probably other solutions, but they would involve fractions or radicals. The one integral solution is just fine.

11 Find the measures of the other two sides of a 30-60-90 right triangle if the shorter leg is 2 inches long. $\approx$ **3.46 inches and 4 inches.**

The longer leg measures $2\sqrt{3} \approx 3.46$ inches, and the hypotenuse is twice the shorter leg, 4 inches.

12 Find the measures of the other two sides of a 30-60-90 right triangle if the longer leg is $7\sqrt{3}$ feet long. **7 feet and 14 feet.**

The shorter leg measures 7 feet, and the hypotenuse measures 14 feet.

13 Find the measures of the other two sides of a 30-60-90 right triangle if the shorter leg is $3\sqrt{3}$ yards long. **9 yards and ≈ 10.39 yards.**

The longer leg is $\sqrt{3}$ times the shorter leg, so it's $\sqrt{3}(3\sqrt{3}) = 3\sqrt{9} = 3(3) = 9$ yards long. The hypotenuse is twice the length of the shorter leg, so it's $2(3\sqrt{3}) = 6\sqrt{3} \approx 10.39$ yards long.

14 Find the measures of the other two sides of a 30-60-90 right triangle if the longer leg is 6 inches long. **≈ 3.46 inches and ≈ 6.93 inches.**

The longer leg is $\sqrt{3}$ times the length of the shorter leg. Let the length of the shorter leg be represented by x. Then $x\sqrt{3} = 6$. Dividing each side by $\sqrt{3}$, you get

$$\frac{x\sqrt{3}}{\sqrt{3}} = \frac{6}{\sqrt{3}}$$

$$x = \frac{6}{\sqrt{3}} \cdot \frac{\sqrt{3}}{\sqrt{3}} = \frac{6\sqrt{3}}{3} = 2\sqrt{3} \approx 3.46 \text{ inches}$$

for the length of the shorter leg. The hypotenuse is twice this or $2(2\sqrt{3}) = 4\sqrt{3} \approx 6.93$ inches.

15 Find the measures of the two legs of the isosceles right triangle that has a hypotenuse measuring 10 feet. **≈ 7.07 feet.**

Let x represent the length of the legs of the triangle. Then

$$x\sqrt{2} = 10$$

$$\frac{x\sqrt{2}}{\sqrt{2}} = \frac{10}{\sqrt{2}}$$

$$x = \frac{10}{\sqrt{2}} \cdot \frac{\sqrt{2}}{\sqrt{2}} = \frac{10\sqrt{2}}{2} = 5\sqrt{2} \approx 7.07 \text{ feet.}$$

16 Find the measures of the three sides of the isosceles right triangle that has an area of 18 square yards. **6 yards, 6 yards, and $6\sqrt{2}$ yards.**

You know that the area of a triangle is half the base times the height. In an isosceles right triangle, the two legs are the base and height. If the area is 18 square yards, then let the legs be represented by x, and

$$A = \frac{1}{2}bh$$

$$18 = \frac{1}{2}x(x) = \frac{1}{2}x^2$$

Multiplying each side by 2 and taking the square root of each side,

$$36 = x^2$$

$$6 = x$$

The two legs are each 6 yards long, which means that the hypotenuse is $6\sqrt{2}$ yards long.

17 Hank is flying a kite and has his entire 100 feet of string out. Hershel is directly under Hank's kite and is standing 28 feet from him. How high is the kite? **96 feet.**

Refer to the figure for the relative positions of the numbers given. Let the unknown height be represented by x. Using the Pythagorean theorem,

$$28^2 + x^2 = 100^2$$
$$784 + x^2 = 10{,}000$$
$$x^2 = 10{,}000 - 784$$
$$x^2 = 9216$$
$$x = 96$$

So the kite is 96 feet up in the air.

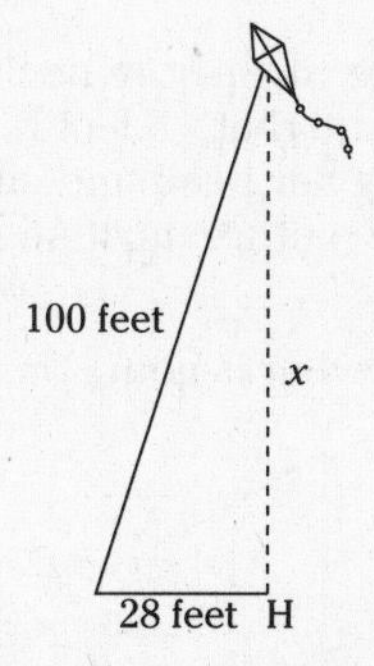

18 How long a fishing rod can you put into a box that's 3 feet by 5 feet by 8 feet? You need to find the diagonal length from one corner to its opposite. ≈ **9.9 feet.**

The fishing rod can be put in the box on the diagonal, so it'll go from the bottom-front corner to the top-back corner on the other side. Refer to the left side of the figure.

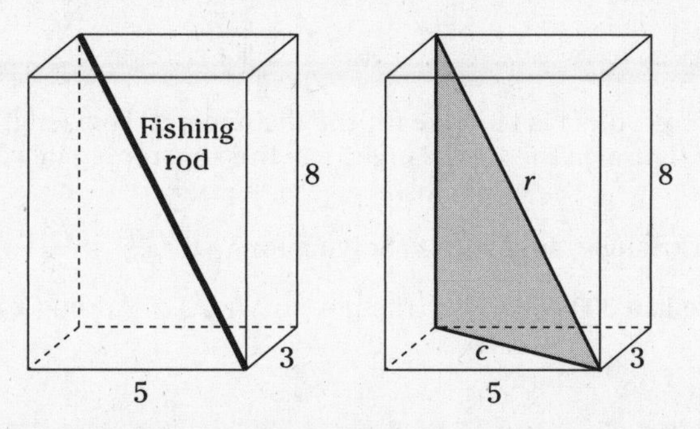

First, find the diagonal across the bottom of the box. This is one of the legs of the right triangle needed to solve the problem. Let the diagonal distance across the bottom be represented by c. Half of the bottom is a right triangle with legs measuring 3 feet and 5 feet. The value of c is the hypotenuse of that right triangle. Solving for c, using the Pythagorean theorem,

$$3^2 + 5^2 = c^2$$
$$9 + 25 = c^2$$
$$34 = c^2$$
$$\sqrt{34} = c$$

So the diagonal across the bottom is $\sqrt{34}$ feet. This is the measure of the bottom leg of the right triangle that's inside the box. The other leg measures 8 feet. Let the measure of the hypotenuse — which is the measure of the rod — be represented by r. Using the two leg measures and the r in the Pythagorean theorem,

$$\left(\sqrt{34}\right)^2 + 8^2 = r^2$$
$$34 + 64 = r^2$$
$$98 = r^2$$
$$r = \sqrt{98} = 7\sqrt{2} \approx 9.9$$

So a fishing rod measuring almost 10 feet will fit in the box. Of course, most fishing rods come apart, but this might be a good old bamboo pole.

19 Gladys is at an art museum and getting a sore neck from looking up and down at the pictures. When she stands in front of a picture that's 6 feet high (its length is 6 feet), she can see 30 degrees above her eye level without tilting her head and she can see 60 degrees below eye level without tilting her head. How far from the picture must she stand to be able to see the whole thing and not move her head? ≈ **2.6 feet.**

Refer to the figure for a sketch of what is going on here.

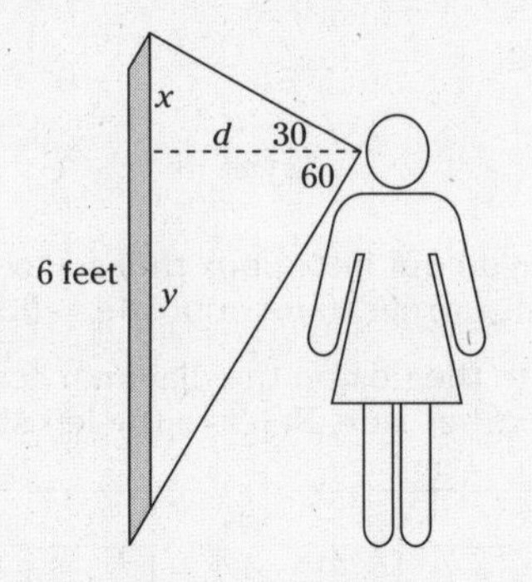

The object of this problem is to solve for the distance, *d*. This length of the picture is the sum of *x* and *y*, from the figure. In the top triangle, *x* is the shorter leg in a 30-60-90 right triangle, and *d* is the

longer leg in that triangle, so $d = \sqrt{3}\,x$. Solving for *x*, $x = \frac{d}{\sqrt{3}} = \frac{\sqrt{3}\,d}{3}$. In the bottom triangle, *y* is the longer leg in a 30-60-90 right triangle, so $y = \sqrt{3}\,d$. Adding *x* and *y* together,

$$x + y = \frac{\sqrt{3}\,d}{3} + \sqrt{3}\,d$$
$$= \frac{\sqrt{3}\,d}{3} + \frac{3\sqrt{3}\,d}{3}$$
$$= \frac{4\sqrt{3}\,d}{3}$$

This sum is equal to 6 feet, so, setting it equal to 6,

$$6 = \frac{4\sqrt{3}\,d}{3}$$
$$18 = 4\sqrt{3}\,d$$
$$d = \frac{18}{4\sqrt{3}} = \frac{9}{2\sqrt{3}} = \frac{9\sqrt{3}}{2\cdot 3} = \frac{3\sqrt{3}}{3} \approx 2.6$$

Gladys needs to stand about 2.6 feet away from the picture.

20 What is the area of a regular *hexagon* (a six-sided polygon with equal sides) when its sides are 4 feet long? (***Hint:*** Subdivide it into six equilateral triangles.) ≈ **41.57 square feet.**

Look at the figure.

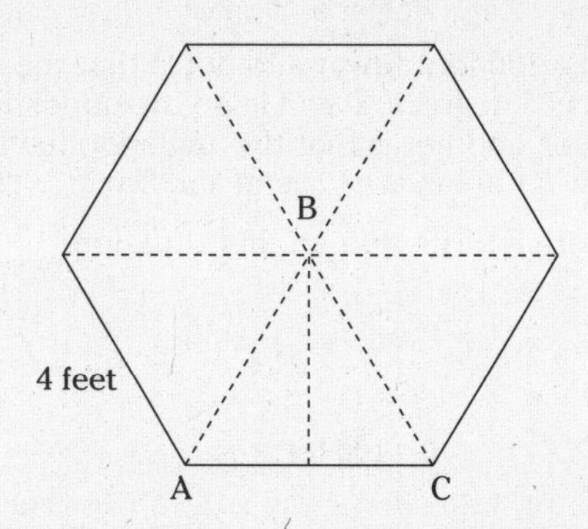

An equilateral triangle has three 60-degree angles. By drawing in the altitude in the bottom triangle, two 30-60-90 right triangles are formed. The lengths of the sides of those two triangles are: 2, 4, and $2\sqrt{3}$. The longer leg is the height of the triangle, and the base is 4 feet. Using the formula for the area of a triangle, the area of one of the equilateral triangles is

$$\begin{aligned} A &= \frac{1}{2}bh \\ &= \frac{1}{2}(4)\left(2\sqrt{3}\right) \\ &= 4\sqrt{3} \text{ square feet} \end{aligned}$$

Because there are six equilateral triangles, multiply this area by 6, and the area of the hexagon is $6\left(4\sqrt{3}\right) = 24\sqrt{3} \approx 41.57$ square feet.

21 Stephanie walks 3 miles east and then 4 miles north. Chelsea joins her, but she walks on the diagonal (the hypotenuse of the right triangle). Chelsea sits and waits while Stephanie walks another 12 miles east and 5 miles north. Again, Chelsea joins her by walking the diagonal. How much farther did Stephanie walk than Chelsea? **6 miles.**

Stephanie is walking the legs of two right triangles, and Chelsea is walking the hypotenuses. Look at the figure, and you'll see what I mean.

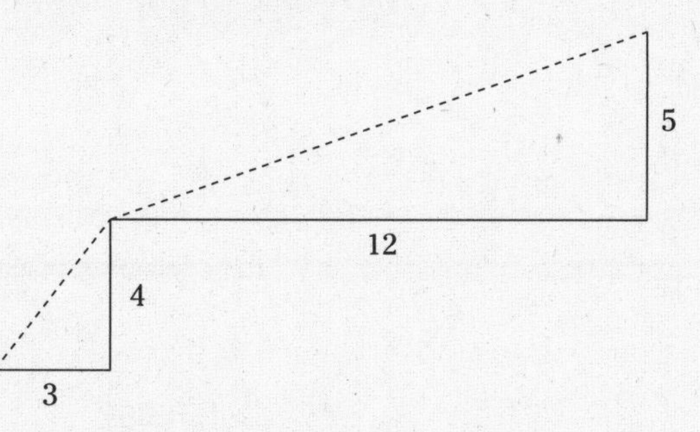

Stephanie walked a total of 24 miles: 3 + 4 + 12 + 5 = 24.

Chelsea walked a total of 18 miles. The first part of her walk was 5 miles. This is the hypotenuse of a 3-4-5 right triangle. The second part of her walk was 13 miles. This is the hypotenuse of a 5-12-13 right triangle.

So Stephanie walked 6 miles farther than Chelsea to get to the same place.

If the numbers hadn't turned out this nicely, you could use the Pythagorean theorem to solve for the missing values in the right triangles.

22 Crystal sights the top of a 100-foot tower and notes that the angle formed from the ground to the top of the tower is 45 degrees. David is on the opposite side of the tower from Crystal, sights the top of the tower, and notes that the angle formed from the ground to the top of the tower is 30 degrees. How far apart are Crystal and David? ≈ **273 feet.**

Look at the figure for more information on this situation.

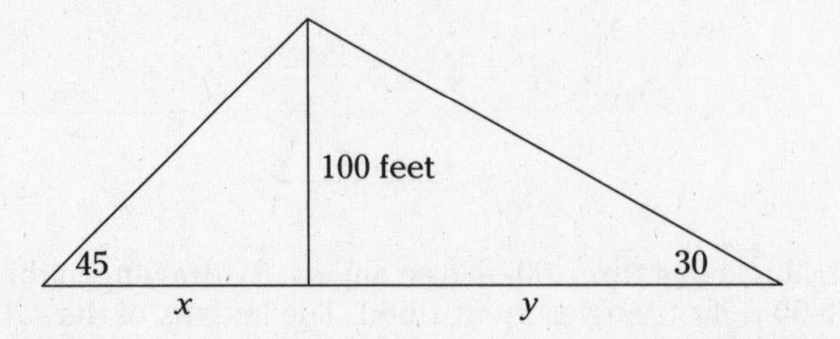

The distance between Crystal and David is the sum of the distances x and y. In the left triangle, x is one of the two equal legs of an isosceles right triangle. The other leg is the 100-foot tower, so x also is 100 feet long. In the right triangle, y is the longer leg of a 30-60-90 right triangle. The shorter leg is 100 feet long, so y is equal to $100\sqrt{3}$. Add the two together, and the distance between the two is $100 + 100\sqrt{3} \approx 273$ feet.

Part II
Trigonometric Functions

In this part . . .

This part may be better known by RSVP: Relating Signs Very Particularly. Here, the signs of numbers and sines of functions are interrelated as the trig functions are described and developed. I cover the six basic trig functions and their respective characteristics here.

Chapter 6

Defining Trig Functions with a Right Triangle

In This Chapter

- Assigning parts of triangles for function values
- Reassigning and coordinating function values
- Using trig functions in applications

The measures of angles, whether they're in degrees or radians, are the input values for trigonometric functions. There are six basic trig functions: sine, cosine, tangent, cotangent, secant, and cosecant. These functions have output values that are real numbers. These output values are the result of operating on angle measures. In this chapter, the functions are defined in terms of the measures of the sides of a right triangle. These definitions are expanded on in Chapter 8, where angles that couldn't be in a right triangle are introduced into the big picture.

The trigonometric functions, when defined in terms of a right triangle, are the ratios between pairs of sides. An acute angle is chosen, and sides are considered with respect to this angle. Use Figure 6-1 in the discussions of the different functions in this chapter.

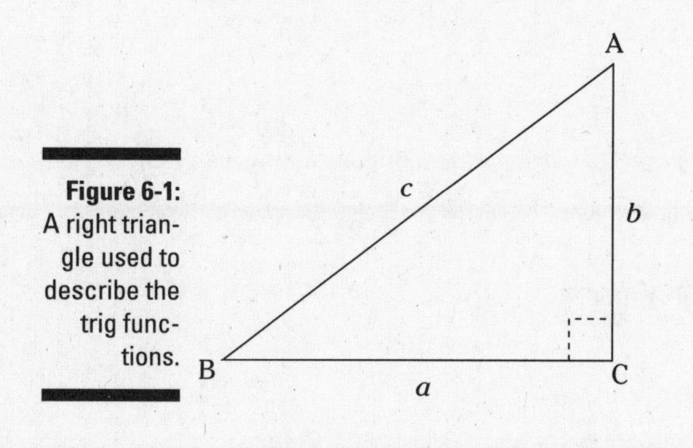

Figure 6-1: A right triangle used to describe the trig functions.

A common way of labeling a triangle is to name the angles with capital letters and the sides opposite the angles with corresponding lowercase letters.

Defining the Sine Function

The *sine* function (not to be confused with *sign* or *sighing*) is defined as being the ratio formed when the length of the side opposite an acute angle in a right triangle is divided by the length of the hypotenuse. In the triangle in Figure 6-1, the sine of angle B is b/c, and the sine of angle A is a/c. The sine is abbreviated *sin*. So the correct mathematical notation for these relationships is $\sin B = \frac{b}{c}$, $\sin A = \frac{a}{c}$.

Q. Use the triangle shown in the figure to determine the sine of angle *A*.

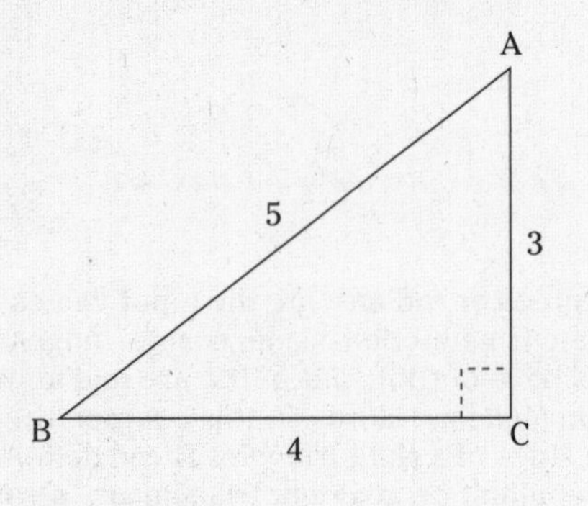

A. Using the triangle, $\sin A = \frac{4}{5}$. It's the opposite side divided by the hypotenuse.

1. Use the triangle in the figure to find sin *B*.

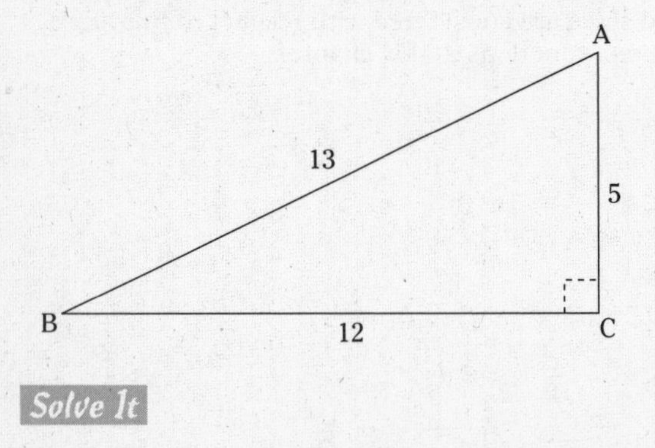

Solve It

2. Use the triangle in the figure to find sin *A*.

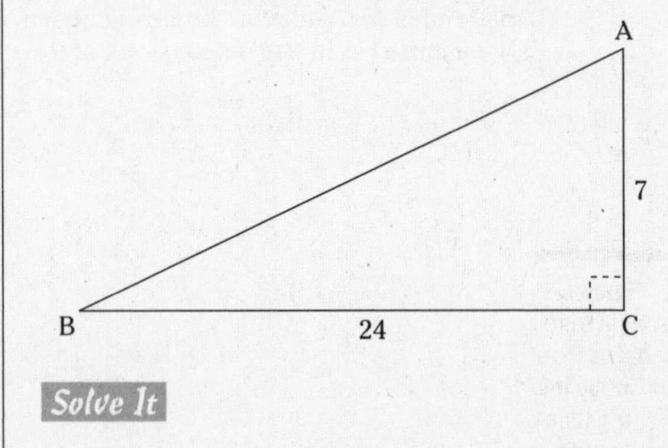

Solve It

Cooperating with the Cosine Function

The *cosine* function is complementary to the sine function, sort of like *cosigning* a loan. The *cosine* is defined as the ratio formed when the length of the side adjacent to an acute angle in a right triangle is divided by the length of the hypotenuse. In Figure 6-1, the cosine of angle *B* is a/c, and the cosine of angle *A* is b/c. The cosine is abbreviated *cos*. The correct mathematical notation for these relationships is $\cos B = \frac{a}{c}$, $\cos A = \frac{b}{c}$.

Q. Use the triangle shown in the figure to determine the cosine of angle *B*.

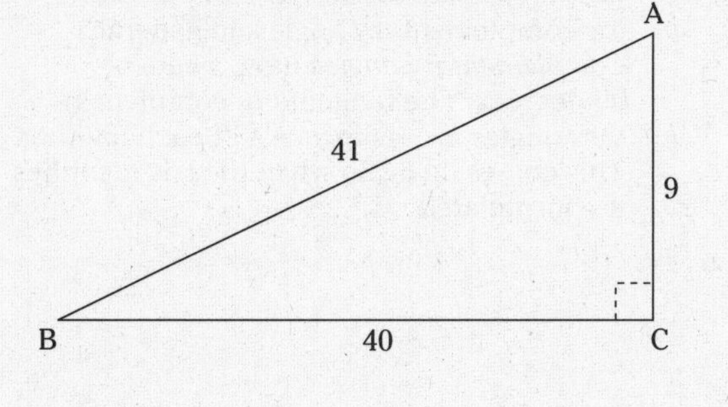

A. The cosine is the length of the adjacent side divided by the length of the hypotenuse, so in this triangle $\cos B = \frac{40}{41}$.

3. Use the triangle in the figure in the example question to find the cosine of angle *A*.

4. Use the triangle in the figure to find the cosine of angle *A*.

A

8

30 degrees

B C

Solve It

Sunning with the Tangent Definition

The *tangent* function could be about soaking up rays (of angles), but it's really more down to earth. The *tangent* is defined as the ratio formed when the length of the side opposite an acute angle in a right triangle is divided by the length of the adjacent side of that same triangle. In Figure 6-1, the tangent of angle B is b/a, and the tangent of angle A is a/b. The tangent is abbreviated *tan.* The correct mathematical notation for these relationships is $\tan B = \frac{b}{a}$, $\tan A = \frac{a}{b}$.

Q. Use the triangle in Figure 6-1 to determine the tangents of angles A and B.

A. $\tan A = \frac{4\sqrt{3}}{4} = \sqrt{3}$ and

$\tan B = \frac{4}{4\sqrt{3}} = \frac{1}{\sqrt{3}} = \frac{\sqrt{3}}{3}$. Do you notice something interesting about these ratios? They're reciprocals of one another. Maybe it's more evident in the definitions, earlier, where the tangents are b/a and a/b. This property holds for all right triangles and for complementary angles, in general. *Complementary angles* have a sum of 90 degrees. The tangents of complementary angles are reciprocals of each another. This comes up again when the trig identities are formulated.

5. Use the triangle in the figure to determine the tangent of angle B.

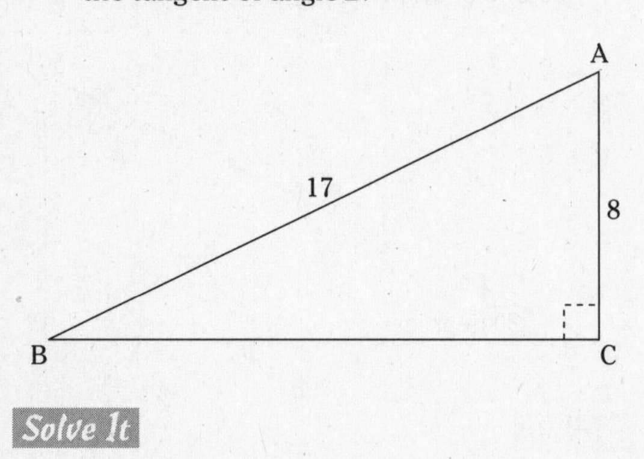

Solve It

6. Find the tangent of angles A and B in the figure.

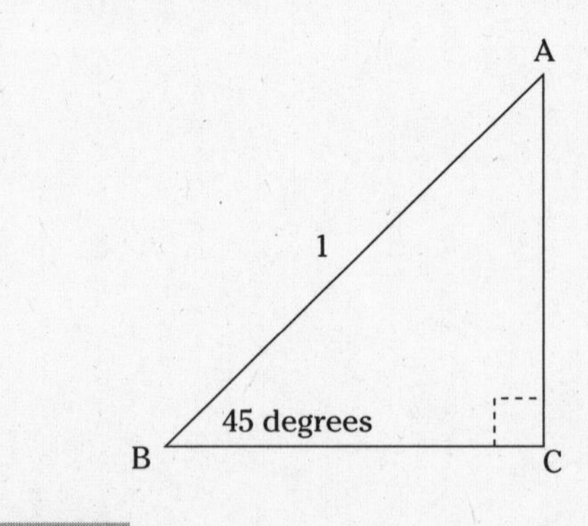

Solve It

Hunting for the Cosecant Definition

The *cosecant* function often makes me feel like I'm telling an insect to look for something: *Go Seek, Ant!* Yes, that's really stretching it, but sometimes it helps to insert some levity when trying to differentiate among all these trig functions.

The *cosecant* function is defined as the ratio formed when the length of the hypotenuse in a right triangle is divided by the length of the opposite side. This function is actually the reciprocal of the sine function, which has the division in the opposite order. In Figure 6-1, the cosecant of angle B is c/b, and the cosecant of angle A is c/a. The cosecant is abbreviated *csc*. The correct mathematical notation for these relationships is $\csc B = \frac{c}{b}$, $\csc A = \frac{c}{a}$.

Q. Use the figure to determine the cosecant of angle B.

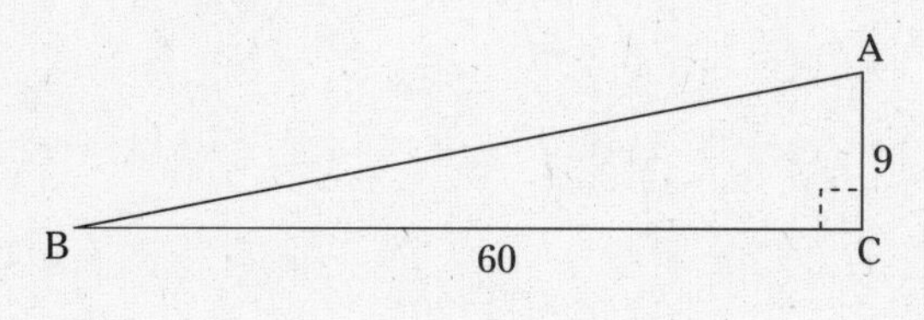

A. I first have to find the length of the hypotenuse. Using the Pythagorean theorem,

$$60^2 + 9^2 = c^2$$
$$3600 + 81 = c^2$$
$$3681 = c^2$$
$$c = 61$$

The hypotenuse is 61 units long. Because the cosecant is the hypotenuse divided by the side opposite angle B, $\csc B = \frac{61}{9}$.

7. Find the cosecant of angle A in the figure in the example question.

Solve It

8. Find the cosecant of the triangle whose sine is $\frac{5}{13}$.

Solve It

9. Use the figure to determine the cosecant of angle *B*.

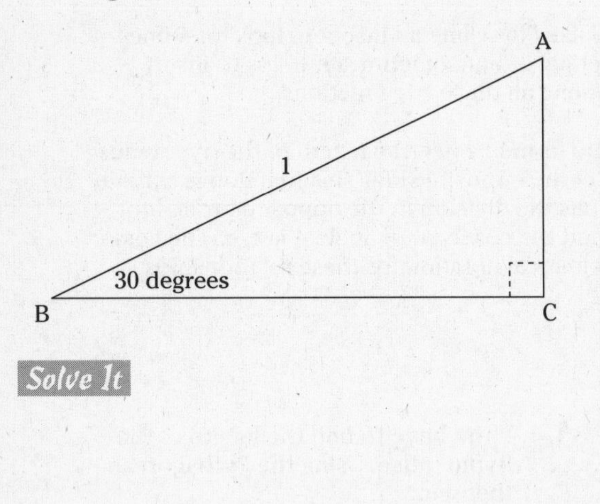

Solve It

10. In a certain 30-60-90 right triangle, the length of the hypotenuse is $2\sqrt{75}$. What is the cosecant of the acute angle opposite the longer leg?

Solve It

Defining the Secant Function

If you keep mixing up the words of songs, you might sing *Oh see can't you say. . . .* The *secant* function is defined as the ratio formed when the length of the hypotenuse in a right triangle is divided by the length of the adjacent side. In Figure 6-1, the secant of angle *B* is *c/a,* and the secant of angle *A* is *c/b.* The secant is abbreviated *sec.* The correct mathematical notation for these relationships is $\sec B = \frac{c}{a}$, $\sec A = \frac{c}{b}$.

Q. If the sides of a right triangle measure 28, 53, and 45, find the secant of the smaller acute angle.

A. The smallest side of a triangle is always opposite the smallest angle, so the side measuring 28 is the opposite side of this angle, and 45 is the adjacent side. The longest side of a right triangle is the hypotenuse, so the ratio of the hypotenuse to the adjacent side in this case is $\frac{53}{45}$.

11. Find the secant of angle A in the triangle in the figure.

A

10

6

B

C

Solve It

12. Find the secant of the acute angles in the triangle shown in the figure.

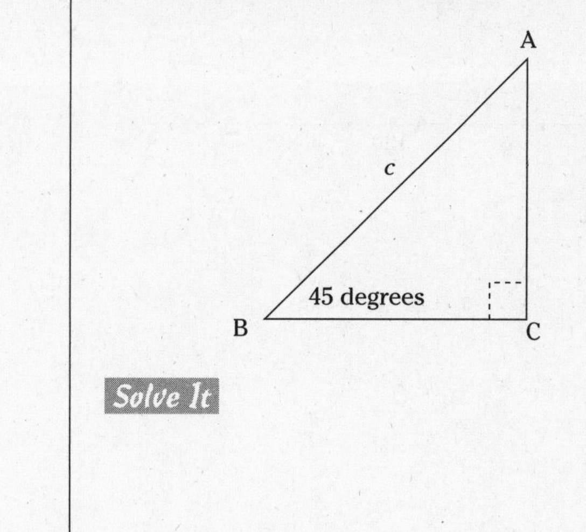

Solve It

Coasting Home with the Cotangent

How is a *cotangent* co- with a tangent? It's because their function values are reciprocals of each another. The *cotangent* is defined as the ratio formed when the length of the side adjacent to an acute angle in a right triangle is divided by the length of the opposite side. In Figure 6-1, the cotangent of angle B is a/b, and the cotangent of angle A is b/a. The cotangent is abbreviated *cot*. The correct mathematical notation for these relationships is $\cot B = \frac{a}{b}$, $\cot A = \frac{b}{a}$.

Q. In a right triangle, the hypotenuse is 1 unit greater than twice the shorter leg, and the longer leg is 1 unit less than twice the shorter leg. What are the cotangents of the two acute angles?

A. The measures of the three sides can be represented by: x, $2x - 1$, and $2x + 1$. The cotangent is adjacent divided by opposite for each acute angle, and the two legs are the x and $2x - 1$. The two ratios formed for the cotangent are: $\frac{x}{2x-1}, \frac{2x-1}{x}$. To find the value of x, use the Pythagorean theorem:

$$x^2 + (2x-1)^2 = (2x+1)^2$$
$$x^2 + 4x^2 - 4x + 1 = 4x^2 + 4x + 1$$

Subtract $4x^2$ and 1 from each side, set the equation equal to 0, and solve for x:

$$x^2 - 4x = 4x$$
$$x^2 - 8x = 0$$
$$x(x-8) = 0$$

The solutions of this equation are 0 or 8. The 0 doesn't make sense, so the 8 is the only solution. This means that the three sides are: $x = 8$, $2x - 1 = 15$, and $2x + 1 = 17$. The cotangent of the smaller of the two acute angles is $\frac{15}{8}$, and the cotangent of the larger angle is $\frac{8}{15}$.

13. Find the cotangent of angle B in the triangle shown in the figure.

A
18
B
120
C

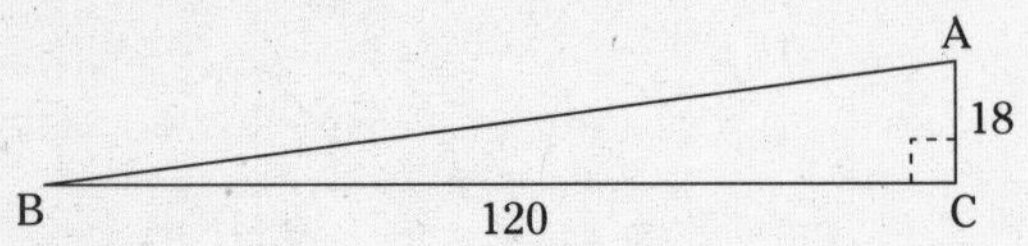

Solve It

14. Compare the cotangents of the smaller of the two acute angles in right triangles with sides: 3, 4, 5; 6, 8, 10; and 15, 20, 25. Do they increase by the amount that the sides are multiplied?

Solve It

15. Find the cotangent of the angle A in the triangle shown in the figure.

A
$5n$
B
$4n$
C

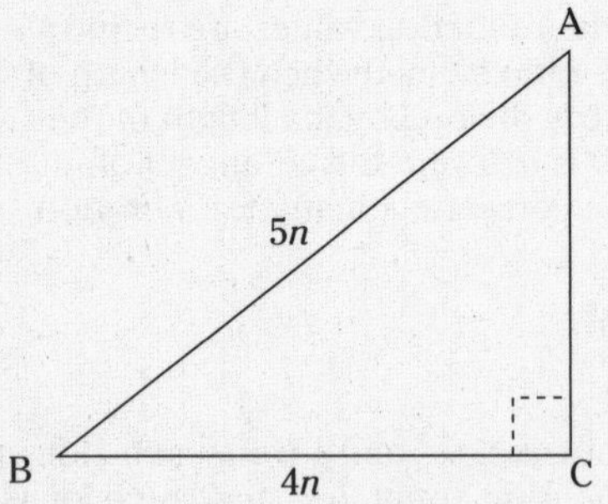

Solve It

16. Find the cotangent of the acute angles in an isosceles right triangle.

Solve It

Establishing Trig Functions for Angles in Special Right Triangles

The trig functions for specific angles are always the same, no matter what size right triangle they're in. The sine of 30 degrees is ½ no matter what. Special right triangles come in very handy when computing the values of the trig functions for their acute angles. The two special right triangles shown in Figure 6-2 have the measures of their sides indicated in general, but, no matter what you let x be (as long as it's a positive number), the different trig functions for the angles will come out the same.

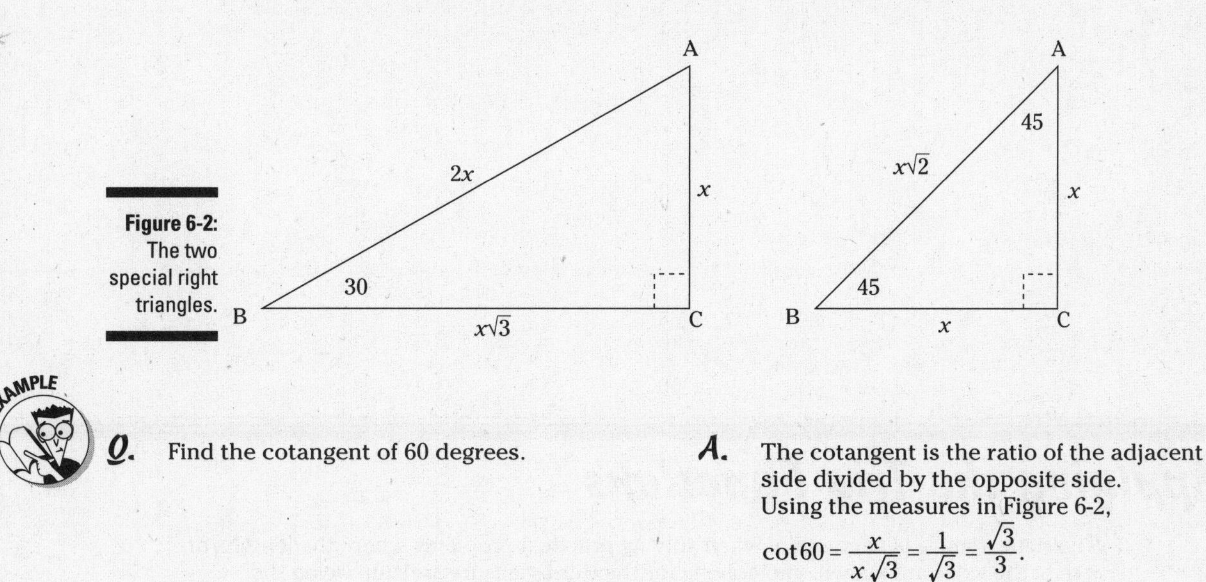

Figure 6-2: The two special right triangles.

EXAMPLE

Q. Find the cotangent of 60 degrees.

A. The cotangent is the ratio of the adjacent side divided by the opposite side. Using the measures in Figure 6-2, $\cot 60 = \frac{x}{x\sqrt{3}} = \frac{1}{\sqrt{3}} = \frac{\sqrt{3}}{3}$.

17. Use Figure 6-2 to find the sine, cosine, and tangent of 30 degrees.

Solve It

18. Use Figure 6-2 to find the sine, cosine, and tangent of 45 degrees.

Solve It

19. Use Figure 6-2 to find the sine, cosine, and tangent of 60 degrees.

Solve It

20. Use Figure 6-2 to find the cosecant, secant, and cotangent of 30 degrees.

Solve It

Applying the Trig Functions

The right triangle is very useful when solving practical problems where the lengths of two of the sides are known, and solving for the third measure ends up being the answer to the problem. Good old Pythagoras comes in handy in those cases.

Using trig functions, problems can be solved when only one side has a known measure. What's needed to complete the problem is the measure of one of the acute angles. The applications here will deal with situations where the known angle is 30 degrees, 45 degrees, or 60 degrees. Other angles can be used if you have a table of values or a scientific calculator, and those work the same way as these problems. Use the function values for the angles from the previous section in this chapter.

EXAMPLE

Q. Doug's golf cart is having trouble. It can't go up an incline that's steeper than 30 degrees. He wants to drive it up a ramp to the top of a 40-foot-high platform. How long a ramp will he need?

A. In the figure, you can see that a 30-60-90 right triangle can be used to solve this. You have the length of the side of the triangle opposite the 30-degree angle. What's needed is the length of the hypotenuse. The sine function is the ratio of the opposite divided by the hypotenuse. The sine of 30 degrees is ½, so write a proportion using this value and the function, and solve for the unknown, represented by the *x*:

$$\sin 30 = \frac{1}{2} = \frac{\text{opposite}}{\text{hypotenuse}}$$

$$\frac{1}{2} = \frac{40}{x}$$

Cross multiply to get $x = 80$. The ramp needs to be 80 feet long.

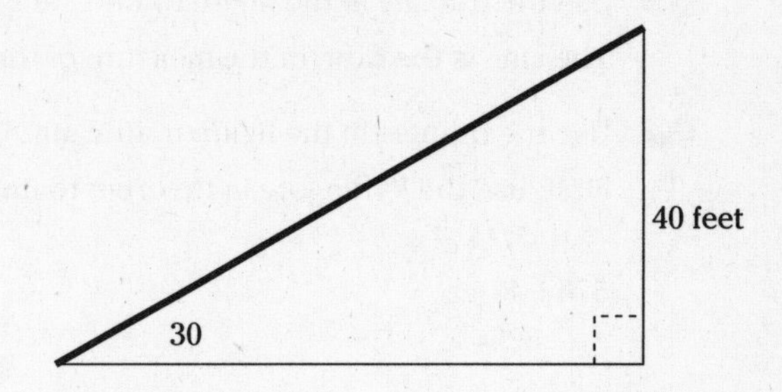

21. Jim and Jane are walking east toward the main road. Jane decides to cut diagonally to the north at an angle of 30 degrees north of east. Jim continues on for the last 20 feet to the road. How far apart are they when they both reach the road? (See the figure for an illustration.)

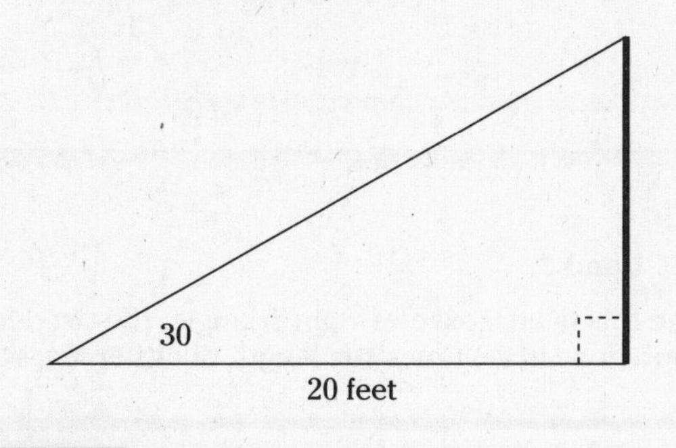

Solve It

22. When Attila leans his ladder against the castle wall, the base of the ladder is 9 feet from the wall, and the angle formed is 60 degrees. How long is the ladder? (See the figure for an illustration.)

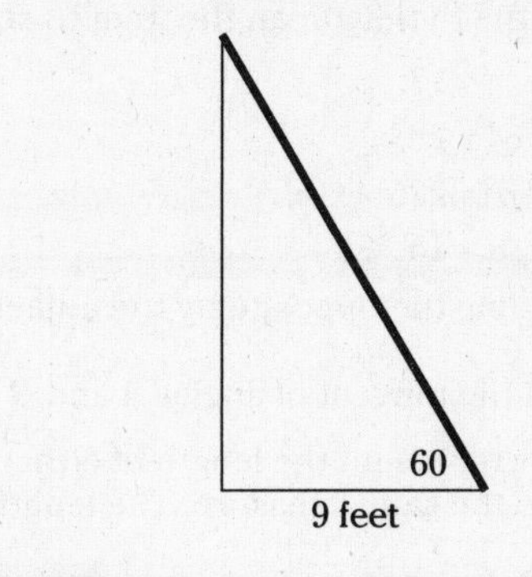

Solve It

Answers to Problems on Defining Trig Functions

The following are the solutions to the practice problems presented earlier in this chapter.

1 Use the triangle in the figure to find sin *B*. $\mathbf{\sin B = \frac{5}{13}}$.

The sine is the ratio of the measure of the opposite side divided by the hypotenuse. $\sin B = \frac{5}{13}$.

2 Use the triangle in the figure to find sin *A*. $\mathbf{\sin A = \frac{24}{25}}$.

First, use the Pythagorean theorem to find the measure of the missing side:

$$\begin{aligned} 24^2 + 7^2 &= c^2 \\ 576 + 49 &= c^2 \\ 625 &= c^2 \\ c &= 25 \end{aligned}$$

The hypotenuse is 25 units long, so $\sin A = \frac{24}{25}$.

3 Use the triangle in the figure in the example question to find the cosine of angle *A*. $\mathbf{\cos A = \frac{9}{41}}$.

The cosine is the ratio of the adjacent side divided by the hypotenuse: $\cos A = \frac{9}{41}$.

4 Use the triangle in the figure to find the cosine of angle *A*. $\mathbf{\frac{1}{2}}$.

In a 30-60-90 right triangle, the length of the shorter leg is half the length of the hypotenuse. Half of 8 is 4; this shorter side is the side adjacent to the 60-degree angle, so the cosine of a 60-degree angle is found with: $\cos 60 = \frac{4}{8} = \frac{1}{2}$.

5 Use the triangle in the figure to determine the tangent of angle *B*. $\mathbf{\frac{8}{15}}$.

The tangent is the length of the opposite side divided by the length of the adjacent side, so first use the Pythagorean theorem to solve for the missing side:

$$\begin{aligned} a^2 + 8^2 &= 17^2 \\ a^2 + 64 &= 289 \\ a^2 &= 225 \\ a &= 15 \end{aligned}$$

Dividing the opposite by the adjacent, $\tan B = \frac{8}{15}$.

6 Find the tangent of angles *A* and *B* in the figure. **1 and 1.**

Let *x* represent the length of either leg. Because this is an isosceles right triangle, the two legs have the same measure. The length of the hypotenuse is $\sqrt{2}$ times the length of either leg, so

$$1 = x\sqrt{2}$$

$$x = \frac{1}{\sqrt{2}} = \frac{\sqrt{2}}{2}$$

The tangent is equal to the opposite divided by the adjacent. They're the same value, so

$$\tan 45 = \frac{\frac{\sqrt{2}}{2}}{\frac{\sqrt{2}}{2}} = 1$$

7 Find the cosecant of angle A in the figure in the example question. $\mathbf{\frac{61}{60}}$.

The cosecant is equal to the hypotenuse divided by the opposite side. First, use the Pythagorean theorem to find the length of the missing side.

$$60^2 + 9^2 = c^2$$
$$3600 + 81 = c^2$$
$$3681 = c^2$$
$$c = 61$$

Then, using the definition of the cosecant, $\csc A = \frac{61}{60}$.

8 Find the cosecant of the triangle whose sine is $\frac{5}{13}$. $\mathbf{\frac{13}{5}}$.

The cosecant is the reciprocal of the sine, so the cosecant must be $\frac{13}{5}$.

9 Use the figure to determine the cosecant of angle B. **2.**

The shorter leg of a 30-60-90 right triangle is half the length of the hypotenuse, so the shorter leg measures ½. Dividing the length of the hypotenuse by the length of the shorter leg (the one opposite the 30-degree angle), $\csc 30 = \frac{1}{\frac{1}{2}} = 2$.

10 In a certain 30-60-90 right triangle, the length of the hypotenuse is $2\sqrt{75}$. What is the cosecant of the acute angle opposite the longer leg? $\mathbf{\csc 60 = \frac{2\sqrt{75}}{15} = \frac{2\sqrt{25}\sqrt{3}}{15} = \frac{2(5)\sqrt{3}}{15} = \frac{2\sqrt{3}}{3}}$.

First, you need find the length of the longer leg. The shorter leg is half the hypotenuse, or $\sqrt{75}$. The longer leg is $\sqrt{3}$ times that length, or $\sqrt{3}\sqrt{75} = \sqrt{225} = 15$. The cosecant of the 60-degree angle is found by dividing the hypotenuse by 15: $\csc 60 = \frac{2\sqrt{75}}{15} = \frac{2\sqrt{25}\sqrt{3}}{15} = \frac{2(5)\sqrt{3}}{15} = \frac{2\sqrt{3}}{3}$.

11 Find the secant of angle A in the triangle in the figure. $\mathbf{\frac{5}{3}}$.

The secant is the length of the hypotenuse divided by the length of the adjacent side, so $\sec A = \frac{10}{6} = \frac{5}{3}$.

12 Find the secant of the acute angles in the triangle shown in the figure. $\sec 45 = \frac{c}{\frac{c}{\sqrt{2}}} = \frac{c}{1} \cdot \frac{\sqrt{2}}{c} = \sqrt{2}$.

The measures of the sides of that isosceles right triangle can be represented by x.

$$x\sqrt{2} = c$$
$$x = \frac{c}{\sqrt{2}}$$

The secant is the hypotenuse divided by the adjacent side (although, here, the two sides, or legs, are the same). This 45-degree angle has $\sec 45 = \frac{c}{\frac{c}{\sqrt{2}}} = \frac{c}{1} \cdot \frac{\sqrt{2}}{c} = \sqrt{2}$.

13 Find the cotangent of angle B in the triangle shown in the figure. $\mathbf{\frac{20}{3}}$.

The cotangent is the adjacent side divided by the opposite side, so $\cot B = \frac{120}{18} = \frac{20}{3}$.

14 Compare the cotangents of the smaller of the two acute angles in right triangles with sides: 3, 4, 5; 6, 8, 10; and 15, 20, 25. Do they increase by the amount that the sides are multiplied? **No.**

The cotangents in each case are equal to $\frac{5}{4}$. They don't change with the lengths of the sides, because the angles in similar triangles are equal.

15 Find the cotangent of the angle A in the triangle shown in the figure. $\mathbf{\frac{3}{4}}$.

First, use the Pythagorean theorem to find the length of the missing side:

$$(4n)^2 + b^2 = (5n)^2$$
$$16n^2 + b^2 = 25n^2$$
$$b^2 = 9n^2$$
$$b = 3n$$

The cotangent is found by dividing the adjacent side by the opposite side: $\cot A = \frac{3n}{4n} = \frac{3}{4}$.

16 Find the cotangent of the acute angles in an isosceles right triangle. **1.**

In an isosceles right triangle, the measures of the two legs are the same. So, for either of the acute angles, the cotangent will be a fraction with numerator and denominator the same. The cotangent is equal to 1.

17 Use Figure 6-2 to find the sine, cosine, and tangent of 30 degrees. $\mathbf{\frac{1}{2}, \frac{\sqrt{3}}{2}}$, **and** $\mathbf{\frac{\sqrt{3}}{3}}$.

$$\sin 30 = \frac{x}{2x} = \frac{1}{2}$$
$$\cos 30 = \frac{x\sqrt{3}}{2x} = \frac{\sqrt{3}}{2}$$
$$\tan 30 = \frac{x}{x\sqrt{3}} = \frac{1}{\sqrt{3}} = \frac{\sqrt{3}}{3}$$

18 Use Figure 6-2 to find the sine, cosine, and tangent of 45 degrees. $\mathbf{\frac{\sqrt{2}}{2}, \frac{\sqrt{2}}{2}}$, **1.**

$$\sin 45 = \frac{x}{x\sqrt{2}} = \frac{1}{\sqrt{2}} = \frac{\sqrt{2}}{2}$$
$$\cos 45 = \frac{x}{x\sqrt{2}} = \frac{1}{\sqrt{2}} = \frac{\sqrt{2}}{2}$$
$$\tan 45 = \frac{x}{x} = 1$$

19 Use Figure 6-2 to find the sine, cosine, and tangent of 60 degrees. $\mathbf{\frac{\sqrt{3}}{2}, \frac{1}{2}, \sqrt{3}}$.

$$\sin 60 = \frac{x\sqrt{3}}{2x} = \frac{\sqrt{3}}{2}$$
$$\cos 60 = \frac{x}{2x} = \frac{1}{2}$$
$$\tan 60 = \frac{x\sqrt{3}}{x} = \sqrt{3}$$

20 Use Figure 6-2 to find the cosecant, secant, and cotangent of 30 degrees. **2,** $\mathbf{\frac{2\sqrt{3}}{3}, \sqrt{3}}$.

$$\csc 30 = \frac{2x}{x} = 2$$
$$\sec 30 = \frac{2x}{x\sqrt{3}} = \frac{2}{\sqrt{3}} = \frac{2\sqrt{3}}{3}$$
$$\cot 30 = \frac{x\sqrt{3}}{x} = \sqrt{3}$$

21 Jim and Jane are walking east toward the main road. Jane decides to cut diagonally to the north at an angle of 30 degrees north of east. Jim continues on for the last 20 feet to the road. How far apart are they when they both reach the road? (See the figure for an illustration.) **≈ 11.5 feet.**

Let x represent the distance between Jim and Jane. This distance is opposite a 30-degree angle. The measure of 20 feet is the adjacent side of the 30-degree angle, so the tangent function should be used. Set up a proportion using the tangent of 30 degrees and the opposite side divided by the adjacent side:

$$\tan 30 = \frac{\sqrt{3}}{3} = \frac{\text{opposite}}{\text{adjacent}} = \frac{x}{20}$$
$$\frac{\sqrt{3}}{3} = \frac{x}{20}$$
$$\frac{20\sqrt{3}}{3} = x$$

This is about 11.5 feet.

22 When Attila leans his ladder against the castle wall, the base of the ladder is 9 feet from the wall, and the angle formed is 60 degrees. How long is the ladder? (See the figure for an illustration.) **18 feet.**

Let x represent the length of the ladder. Use the cosine of 60 degrees to solve this:

$$\cos 60 = \frac{1}{2} = \frac{\text{adjacent}}{\text{hypotenuse}} = \frac{9}{x}$$
$$\frac{1}{2} = \frac{9}{x}$$
$$x = 18$$

The ladder is 18 feet long.

Chapter 7

Discussing Properties of the Trig Functions

In This Chapter

- Defining functions and inverses
- Describing domains and ranges of trig functions
- Determining exact values for functions

The trig functions act like other mathematical functions, because they follow the rule that there can be only one output for every input value. What makes the trig functions unique is that the input values are all angle measures and, also, that the trig functions are *periodic* — the function values repeat over and over and over in a predictable pattern. The trig functions can be used to model natural occurrences that repeat over and over such as weather, planet revolutions, and seasonal sales.

Defining a Function and Its Inverse

A function is a relationship between input and output values in which there's exactly one output for every input in the function's domain. The *domain* consists of all the possible input values. Some functions have inverses. An *inverse* reverses the result of the function operation and tells you what you started with or what was input to get that. Not all functions have inverses. The trig functions technically have inverses only for small parts of their domains. The inverse trig functions are thoroughly covered in Chapter 12.

Q. The function $f(x)=\frac{x^5-3}{2}$ has an inverse: $f^{-1}(x)=\sqrt[5]{2x+3}$. Demonstrate that these two functions are inverses by inputting 2 into the first function and the result of that input into the second function.

A. Showing how inverses work, I'll input 2 for x, $f(2)=\frac{2^5-3}{2}=\frac{32-3}{2}=\frac{29}{2}=14.5$. Now, putting that result into the inverse function, $f^{-1}(14.5)=\sqrt[5]{2(14.5)+3}=\sqrt[5]{29+3}=\sqrt[5]{32}=2$. I'm right back where I started. This example is for just one number and doesn't prove the inverse for all input numbers.

Q. Prove that the functions in the first example are inverses of one another for all values in the domain. (***Hint:*** Input the entire inverse function into the original function, and show that the result is always x, the original input.)

A. Letting the second function be the input into the first, $f\left(f^{-1}\right)=f\left(\sqrt[5]{2x+3}\right)=$ $\frac{\left(\sqrt[5]{2x+3}\right)^5-3}{2}=\frac{2x+3-3}{2}=\frac{2x}{2}=x$. And then, to finish the problem properly, you should also input the entire original function into the input function, too.

$$\begin{aligned}f^{-1}(f)&=f^{-1}\left(\frac{x^5-3}{2}\right)\\&=\sqrt[5]{2\left(\frac{x^5-3}{2}\right)+3}\\&=\sqrt[5]{x^5-3+3}\\&=\sqrt[5]{x^5}\\&=x\end{aligned}$$

1. Show that the function $f(x) = 2x + 1$ has an inverse that's $f^{-1}(x)=\frac{x-1}{2}$.

Solve It

2. Show that the function $f(x)=\sqrt[3]{x-3}+2$ has an inverse that's $f^{-1}(x)=(x-2)^3+3$.

Solve It

Deciding on the Domains

The *domain* of a function consists of all the input values that the function has. The domains of the sine and cosine functions aren't restricted — they can be anything. But the domains of the other functions have omissions or values that are skipped. Here's a rundown on the functions and their domains:

- **Sine:** All angles; all real numbers.
- **Cosine:** All angles; all real numbers.
- **Tangent:** All angles except 90 degrees, 270 degrees, and odd multiples of those angles; all real numbers except $\frac{\pi}{2}, \frac{3\pi}{2}$, or odd multiples of those values.
- **Cotangent:** All angles except 0 degrees, 180 degrees, and multiples of 180; all real numbers except multiples of π.
- **Secant:** All angles except 90 degrees, 270 degrees, and odd multiples of those angles; all real numbers except $\frac{\pi}{2}, \frac{3\pi}{2}$, or odd multiples of those values.
- **Cosecant:** All angles except 0 degrees, 180 degrees, and multiples of 180; all real numbers except multiples of π.

Q. Which of the trig functions have all angles between 0 and 180 degrees in their domains?

A. The sine, cosine, cosecant, and cotangent have the angles between 0 and 180 in their domains. The sine and cosine also include 0 and 180. The cosecant and cotangent don't include those two values. The tangent and secant were excluded from this answer, because neither includes the 90-degree angle in its domain.

3. Which functions have all the angles between 0 and 90 degrees in their domains?

Solve It

4. Which functions have all the angles between 90 and 270 degrees in their domains?

5. Which functions have an angle measuring 90 degrees in their domains?

Solve It

6. Which functions have an angle measuring 180 degrees in their domains?

Solve It

7. Which functions have $\frac{3\pi}{2}$ in their domains?

Solve It

8. Which functions have 0 in their domains?

Solve It

Reaching Out for the Ranges

The *range* of a function consists of all the values that are a result of performing the function on the input. It's all the output values. The following are the ranges of the trigonometric functions:

- **Sine:** All real numbers between –1 and 1, including those two values
- **Cosine:** All real numbers between –1 and 1, including those two values
- **Tangent:** All real numbers
- **Cotangent:** All real numbers
- **Secant:** All real numbers 1 and greater, and all real numbers –1 and smaller
- **Cosecant:** All real numbers 1 and greater, and all real numbers –1 and smaller

As you can see, the sine and cosine are caught between –1 and 1, the secant and cosecant don't have any range values between –1 and 1, and the tangent and cotangent are all over the place.

Q. Which of the functions have a range that includes all the positive numbers?

A. This can be only the tangent and cotangent. The others have ranges that include some positive numbers, but not all of them.

9. Which functions have a range that includes numbers greater than 1?

10. Which functions have a range that includes numbers between –1 and 0?

Closing In on Exact Values

An *exact value* is a number that isn't rounded off or truncated. It's exactly the correct value. For instance, the fraction $\frac{5}{8}$ is exact, and its decimal equivalent, 0.625 is exact. The fraction $\frac{1}{6}$ is exact, but it has a repeating decimal, 0.1666 . . . that keeps going on forever. You can round this decimal off to 0.1667 or some other such value, but that isn't exact; it's an approximation — even though the rounded version is more convenient.

The exact values for trig functions are the ones that are written as integers or rational numbers or radicals. These values are often preferred for very exacting computations. When someone is building a bridge, you'd prefer she be *exact.* The exact values of the most commonly used angles in trig functions are those for the angles found in the special right triangles.

Two more angles are going to be added here, 0 degrees and 90 degrees. An angle of 0 degrees has an opposite side that measures 0. The adjacent side and hypotenuse are the same. Just for convenience, let the adjacent side and hypotenuse have measures of x. Then $\sin 0 = \frac{\text{opposite}}{\text{hypotenuse}} = \frac{0}{x} = 0$, $\cos 0 = \frac{\text{adjacent}}{\text{hypotenuse}} = \frac{x}{x} = 1$, and $\tan 0 = \frac{\text{opposite}}{\text{adjacent}} = \frac{0}{x} = 0$. The 90-degree angle is a little harder to explain, because you can't have two 90-degree angles in one triangle, but if you let the opposite side and the hypotenuse be the same measure and let the adjacent side be 0, you'll get the function values for 90 degrees: $\sin 90 = \frac{\text{opposite}}{\text{hypotenuse}} = \frac{x}{x} = 1$, $\cos 90 = \frac{\text{adjacent}}{\text{hypotenuse}} = \frac{0}{x} = 0$, and $\tan 90 = \frac{\text{opposite}}{\text{adjacent}} = \frac{x}{0}$. This last one has no value — but that's consistent with the domain of the tangent function, which doesn't include 90 degrees.

Q. What is the exact value of the sine of 30 degrees?

A. The exact value is $\frac{1}{2}$.

Q. What is the exact value of the tangent of 45 degrees?

A. The exact value is 1.

11. Fill in the exact values for the cosine function in Table 7-1.

Table 7-1 Exact Values for Cosine

Degrees	0	30	45	60	90
Cosine					

12. Fill in the exact values for the tangent function in Table 7-2.

Table 7-2 Exact Values for Tangent

Degrees	0	30	45	60	90
Tangent					

Solve It

13. Fill in the exact values for the sine function in Table 7-3.

Table 7-3 Exact Values for Sine

Degrees	0	30	45	60	90
Sine					

Solve It

Determining Exact Values for All Functions

As handy as the special right triangles are for finding exact values, you sometimes need a quick trick for finding those values and using them in a problem. I start by creating a table with the sine and cosine of angles measuring 0, 30, 45, 60, and 90 degrees (see Table 7-5).

Table 7-5 Sine and Cosine Function Values

Degrees	0	30	45	60	90
Sine	0	$\frac{1}{2}$	$\frac{\sqrt{2}}{2}$	$\frac{\sqrt{3}}{2}$	1
Cosine	1	$\frac{\sqrt{3}}{2}$	$\frac{\sqrt{2}}{2}$	$\frac{1}{2}$	0

The value for the tangent of each angle is found by dividing the sine by the cosine. The cotangent of each angle is the reciprocal of the tangent. The secant is the reciprocal of the cosine. And the cosecant is the reciprocal of the sine. These are all derived from the definitions of the different functions in the right triangle — with a little hand-waving for the angles measuring 0 and 90.

Q. How can you determine the exact value of the secant of 60 degrees using a value in Table 7-5?

A. The secant is the reciprocal of the cosine. Just flip the 1/2, and you get that the secant of 60 degrees is $\frac{2}{1}$ which is, of course, 2.

Q. How can you determine the exact value of the cotangent of 0 degrees using a value in Table 7-5?

A. The cotangent is the reciprocal of the tangent. The tangent of 0 degrees is 0, and that number doesn't have a reciprocal (you can't divide by 0). So the cotangent doesn't have any value, let alone an exact one, for 0 degrees.

14. Fill in the exact values for the tangent function in Table 7-6, using sine divided by cosine.

Table 7-6 Exact Values for Tangent

Degrees	0	30	45	60	90
Tangent					

Solve It

15. Fill in the exact values for the cotangent function in Table 7-7, using the reciprocal of the tangent. (***Note:*** For 90 degrees, use cosine over sine.)

Table 7-7 Exact Values for Cotangent

Degrees	0	30	45	60	90
Cotangent					

16. Fill in the exact values for the secant function in Table 7-8, using the reciprocal of the cosine.

Table 7-8 Exact Values for Secant

Degrees	0	30	45	60	90
Secant					

Solve It

17. Fill in the exact values for the cosecant function in Table 7-9, using the reciprocal of the sine.

Table 7-9 Exact Values for Cosecant

Degrees	0	30	45	60	90
Cosecant					

Solve It

Answers to Problems in Properties of Trig Functions

The following are solutions to the practice problems presented earlier in this chapter.

1 Show that the function $f(x) = 2x + 1$ has an inverse that's $f^{-1}(x) = \frac{x-1}{2}$.

First insert the inverse function into the original function as an input value: $f\left(\frac{x-1}{2}\right) = 2\left(\frac{x-1}{2}\right) + 1 = \cancel{2}\left(\frac{x-1}{\cancel{2}}\right) + 1 = x - 1 + 1 = x$. The result is x, which is what is needed.

Now insert the original function into the inverse function as input to see if you get x this time, too: $f^{-1}(2x+1) = \frac{2x+1-1}{2} = \frac{2x}{2} = x$

2 Show that the function $f(x) = \sqrt[3]{x-3} + 2$ has an inverse that's $f^{-1}(x) = (x-2)^3 + 3$.

First insert the inverse function into the original function as an input value: $f\left((x-2)^3 + 3\right) = \sqrt[3]{(x-2)^3 + 3 - 3} + 2 = \sqrt[3]{(x-2)^3} + 2 = x - 2 + 2 = x$. Then insert the original function into the inverse function as input: $f^{-1}\left(\sqrt[3]{x-3} + 2\right) = \left(\sqrt[3]{x-3} + 2 - 2\right)^3 + 3 = \left(\sqrt[3]{x-3}\right)^3 + 3 = x - 3 + 3 = x$. They both have a result x,
which is what is needed.

3 Which functions have all the angles between 0 and 90 degrees in their domains? **Sine, cosine, tangent, cotangent, secant, and cosecant.**

All six trig functions have the angles between 0 and 90 degrees in their domains. The tangent and secant have domains that don't include 90 degrees. The cotangent and cosecant have domains that don't include 0 degrees.

4 Which functions have all the angles between 90 and 270 degrees in their domains? **Sine, cosine, tangent, and secant.**

The sine, cosine, tangent, and secant have all the angles between 90 and 270 in their domains. The tangent and secant have domains that don't include those two endpoints, 90 and 270, though.

5 Which functions have an angle measuring 90 degrees in their domains? **Sine, cosine, cotangent, and cosecant.**

6 Which functions have an angle measuring 180 degrees in their domains? **Sine, cosine, tangent, and secant.**

7 Which functions have $\frac{3\pi}{2}$ in their domains? **Sine, cosine, cotangent, and cosecant.**

8 Which functions have 0 in their domains? **Sine, cosine, tangent, and secant.**

9 Which functions have a range that includes numbers greater than 1? **Tangent, cotangent, secant, and cosecant.**

10 Which functions have a range that includes numbers between –1 and 0? **Sine, cosine, tangent, and cotangent.**

11 Fill in the exact values for the cosine function in Table 7-1. **1; $\frac{\sqrt{3}}{2}$; $\frac{\sqrt{2}}{2}$; $\frac{1}{2}$; 0.**

12 Fill in the exact values for the tangent function in Table 7-2. **0; $\frac{\sqrt{3}}{3}$; 1; $\sqrt{3}$; undefined (no value).**

13 Fill in the exact values for the sine function in Table 7-3. **0; $\frac{1}{2}$; $\frac{\sqrt{2}}{2}$; $\frac{\sqrt{3}}{2}$; 1.**

14 Fill in the exact values for the tangent function in Table 7-6, using sine divided by cosine.

$$\frac{0}{1} = 0;$$

$$\frac{\frac{1}{2}}{\frac{\sqrt{3}}{2}} = \frac{1}{2} \cdot \frac{2}{\sqrt{3}} = \frac{1}{\sqrt{3}} = \frac{\sqrt{3}}{3};$$

$$\frac{\frac{\sqrt{2}}{2}}{\frac{\sqrt{2}}{2}} = 1;$$

$$\frac{\frac{\sqrt{3}}{2}}{\frac{1}{2}} = \frac{\sqrt{3}}{2} \cdot \frac{2}{1} = \sqrt{3};$$

$$\frac{1}{0} \text{ undefined (no value)}$$

15 Fill in the exact values for the cotangent function in Table 7-7, using the reciprocal of the tangent. (***Note:*** For 90 degrees, use cosine over sine.)

$$\frac{1}{0} \text{ undefined (no value);}$$

$$\frac{1}{\frac{\sqrt{3}}{3}} = \frac{3}{\sqrt{3}} = \sqrt{3};$$

$$\frac{1}{1} = 1;$$

$$\frac{1}{\sqrt{3}} = \frac{\sqrt{3}}{3};$$

$$\frac{0}{1} = 0$$

16 Fill in the exact values for the secant function in Table 7-8, using the reciprocal of the cosine.

$$\frac{1}{1} = 1;$$

$$\frac{1}{\frac{\sqrt{3}}{2}} = \frac{2}{\sqrt{3}} = \frac{2\sqrt{3}}{3};$$

$$\frac{1}{\frac{\sqrt{2}}{2}} = \frac{2}{\sqrt{2}} = \sqrt{2};$$

$$\frac{1}{\frac{1}{2}} = 2;$$

$$\frac{1}{0} \text{ undefined (no value)}$$

17 Fill in the exact values for the cosecant function in Table 7-9, using the reciprocal of the sine.

$\frac{1}{0}$ **undefined (no value);**

$\frac{1}{\frac{1}{2}} = 2;$

$\frac{1}{\frac{\sqrt{2}}{2}} = \frac{2}{\sqrt{2}} = \sqrt{2};$

$\frac{1}{\frac{\sqrt{3}}{2}} = \frac{2}{\sqrt{3}} = \frac{2\sqrt{3}}{3};$

$\frac{1}{1} = 1$

Chapter 8

Going Full Circle with the Circular Functions

In This Chapter

- Investigating the unit circle
- Using reference angles and quadrant signs
- Finding trig functions for all angles

Describing the trig functions using a right triangle makes sense and fosters good understanding of what the trig functions are and do. But there's a much bigger picture to consider when dealing with angles and the trig functions.

Instead of creating function values for every angle there is — positive or negative, one or more full rotations — a better system exists. All angles can be related to other angles that are within one full rotation. They're related by their shared terminal side. You just have to focus on the angles within one rotation — those between 0 and 360 degrees.

Finding Points on the Unit Circle

The *unit circle* is a circle that has a radius of 1 unit. It's most often found with its center at the origin. The equation of this unit circle is $x^2 + y^2 = 1$. Coordinates of the points on this circle satisfy the equation. For instance, the point $\left(-\frac{1}{2}, \frac{\sqrt{3}}{2}\right)$ is on the unit circle. Here's what it looks like to substitute the coordinates into the equation: $\left(-\frac{1}{2}\right)^2 + \left(\frac{\sqrt{3}}{2}\right)^2 = \frac{1}{4} + \frac{3}{4} = 1$.

The sine and cosine of angles measuring 30, 45, and 60 degrees show up as coordinates of points on the unit circle. For example, the following points lie on the unit circle: $\left(\frac{1}{2}, -\frac{\sqrt{3}}{2}\right), \left(\frac{\sqrt{3}}{2}, \frac{1}{2}\right), \left(-\frac{\sqrt{2}}{2}, \frac{\sqrt{2}}{2}\right), \left(-\frac{\sqrt{2}}{2}, -\frac{\sqrt{2}}{2}\right)$. In every case, the sum of the squares of the coordinates is equal to 1.

What about all the other points on the unit circle? There are an infinite number of them. You can choose any number between 0 and 1 — the coordinates can't have values greater than 1.

Q. Let $x = \frac{9}{41}$ and solve for the other coordinate of the point on the unit circle.

A. $y = \frac{40}{41}$ or $y = -\frac{40}{41}$. Substituting that into the equation for the unit circle, and solving for y,

$$\left(\frac{9}{41}\right)^2 + y^2 = 1$$

$$\frac{81}{1681} + y^2 = 1$$

$$y^2 = 1 - \frac{81}{1681} = \frac{1600}{1681}$$

$$y = \pm\frac{40}{41}$$

There are two possible solutions, because one point in the first quadrant and the other in the fourth quadrant have that same x coordinate: $\left(\frac{9}{41}, \frac{40}{41}\right), \left(\frac{9}{41}, -\frac{40}{41}\right)$.

Q. Let $y = 1$ and solve for the other coordinate of the point on the unit circle.

A. $x = 0$. The point (0,1) lies on the y-axis.

1. Fill in the missing coordinates on the unit circle, represented by the letters.

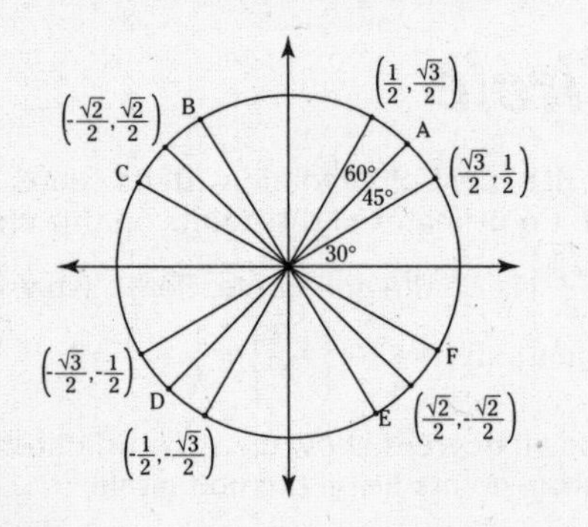

Solve It

2. Fill in the coordinates of the points on the unit circle that lie on the axes. They're represented by the letters A, B, C, and D.

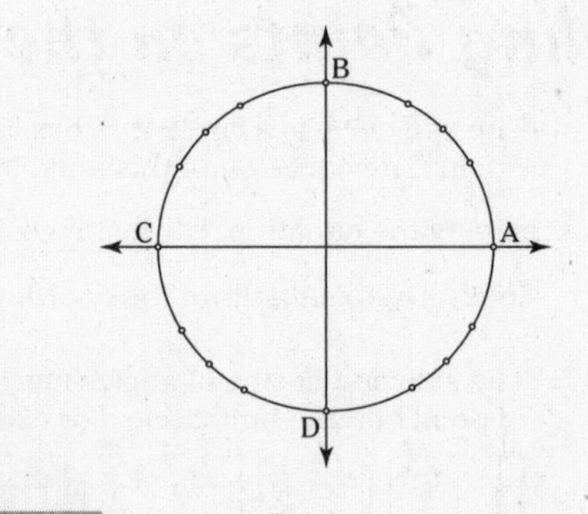

Solve It

3. Find the y coordinate of the point on the unit circle if $x = \frac{12}{13}$, and the point lies in Quadrant I.

Solve It

4. Find the x coordinate of the point on the unit circle if $y = \frac{7}{25}$, and the point lies in Quadrant II.

Solve It

5. Find the y coordinate of the point on the unit circle if $x = -\frac{3}{5}$, and the point lies in Quadrant III.

Solve It

6. Find the x coordinate of the point on the unit circle if $y = -\frac{8}{17}$, and the point lies in Quadrant IV.

7. Find the y coordinate of the point on the unit circle if $x = \frac{1}{3}$, and the point lies in Quadrant I.

Solve It

8. Find the x coordinate of the point on the unit circle if $y = -\frac{2}{7}$, and the point lies in Quadrant III.

Solve It

Determining Reference Angles

The values of the trigonometric functions can be related to the ratios of the sides of a right triangle. Imagine a right triangle sitting in Quadrant II, with a negative value assigned to the adjacent side. Or try Quadrant III, assigning negative values to both the opposite and adjacent sides. This concept is used to assign values to the trig functions that are greater than 90 degrees. You just have to figure out which triangle in which quadrant is to be used. Hence, *reference angles:* The reference angle for a particular angle helps determine the values of the trig functions. Table 8-1 shows you how to determine reference angles.

Table 8-1 Determining the Values of Reference Angles

Angle Measure	*Reference Angle*	*Angle Measure*	*Reference Angle*
$0 < x < 90$	x	$0 < x < \frac{\pi}{2}$	x
$90 < x < 180$	$180 - x$	$\frac{\pi}{2} < x < \pi$	$\pi - x$
$180 < x < 270$	$x - 180$	$\pi < x < \frac{3\pi}{2}$	$x - \pi$
$270 < x < 360$	$360 - x$	$\frac{3\pi}{2} < x < 2\pi$	$2\pi - x$

Q. Find the reference angle for 300 degrees.

A. The measure of 300 degrees lies between 270 and 360. Subtract 360 – 300 = 60 degrees.

Q. Find the reference angle for $\frac{5\pi}{6}$.

A. This angle lies between $\frac{\pi}{2}$ and π. Subtract $\pi - \frac{5\pi}{6} = \frac{6\pi}{6} - \frac{5\pi}{6} = \frac{\pi}{6}$.

9. Find the reference angle for 120 degrees.

Solve It

10. Find the reference angle for 315 degrees.

Solve It

11. Find the reference angle for 210 degrees.

Solve It

12. Find the reference angle for $\frac{5\pi}{4}$.

Solve It

13. Find the reference angle for $\frac{5\pi}{6}$.

Solve It

14. Find the reference angle for $\frac{11\pi}{3}$.

Solve It

Assigning the Signs of Functions by Quadrant

The position of the terminal side of an angle in standard position determines the signs of the different trig functions for the angle. Think of a terminal side as being the hypotenuse of a right triangle. A right triangle in Quadrant III would be assigned a negative opposite side and a negative adjacent side. The way that the different ratios are put together creates either positive or negative results. Instead of trying to piece together the quadrant and the ratio and the division result, you can use a quick, easy method to remember the signs of the trig functions based on the terminal side of the angle. Look at Figure 8-1.

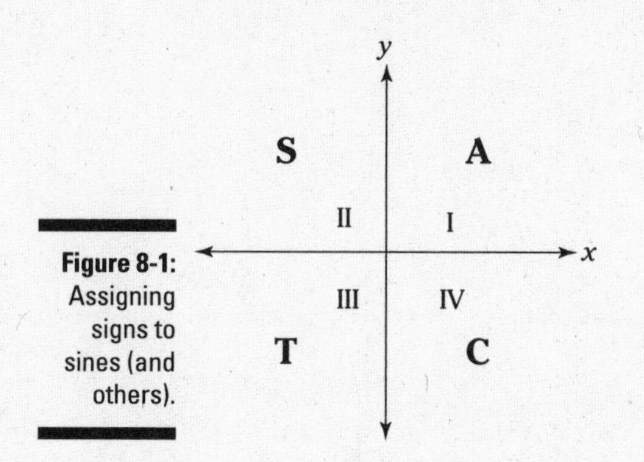

Figure 8-1: Assigning signs to sines (and others).

The A, S, T, and C stand for:

- **A:** All angles
- **S:** Sine (and its reciprocal, cosecant)
- **T:** Tangent (and its reciprocal, cotangent)
- **C:** Cosine (and its reciprocal, secant)

The letters indicate which of the functions is *positive* in the particular quadrant. In Quadrant III, for example, only the tangent and cotangent are positive — all the other functions are negative. I was taught that the letters stand for: *All Students Take Calculus.* You can make up your own saying, if you prefer.

Q. Which functions are positive for an angle measuring 280 degrees?

A. An angle measuring 280 degrees has a terminal side lying in Quadrant IV. That means that the cosine and secant are positive, and all the rest are negative.

Q. Which functions are positive for an angle measuring $\frac{3\pi}{4}$?

A. This angle has a terminal side lying in Quadrant II. The sine and cosecant are the only positive functions for this angle.

15. Which functions are positive for an angle measuring 105 degrees?

16. Which functions are positive for an angle measuring 330 degrees?

17. Which functions are positive for an angle measuring $\frac{4\pi}{3}$?

Solve It

18. Which functions are positive for an angle measuring $\frac{11\pi}{6}$?

19. Which functions are positive for an angle measuring $\frac{7\pi}{3}$?

Solve It

20. Which functions are positive for an angle measuring 927 degrees?

Solve It

Figuring Out Trig Functions around the Clock

Using reference angles and signs of angles in the different quadrants, you can determine the values of trig functions of many angles. The best angles to start with are those that are multiples of 30, 45, and 60 degrees.

To find the value of the trig function:

1. **Determine the quadrant in which the terminal side of the angle lies, and find the sign of the function in that quadrant.**
2. **Determine the reference angle.**
3. **Use the chart on the Cheat Sheet (if you don't have them memorized) that has the exact values of many angles between 0 and 90 degrees to find the function value.**
4. **Apply the sign of the function to the numerical value.**

Q. Find the value of cos 135 degrees.

A. The terminal side is in Quadrant II; the cosine is negative in Quadrant II. Find the reference angle by subtracting 180 – 135 = 45 degrees. The cosine of 45 degrees is $\frac{\sqrt{2}}{2}$. Using the negative, the answer is $-\frac{\sqrt{2}}{2}$.

Q. Find the value of $\csc\frac{16\pi}{3}$.

A. This angle is equal to two full revolutions plus $\frac{4\pi}{3}$. I got this by subtracting, $\frac{16\pi}{3} - 4\pi = \frac{16\pi}{3} - \frac{12\pi}{3} = \frac{4\pi}{3}$. The terminal side is in Quadrant III; the cosecant is negative in Quadrant III. The reference angle is $\frac{4\pi}{3} - \pi = \frac{\pi}{3}$. The cosecant of $\frac{\pi}{3}$ is $\frac{2\sqrt{3}}{3}$. Using the negative, the answer is $-\frac{2\sqrt{3}}{3}$.

21. Find sin 315.

Solve It

22. Find cos 240.

Solve It

23. Find tan $\frac{7\pi}{6}$.

Solve It

24. Find csc $\frac{5\pi}{6}$.

Solve It

Answers to Problems in Going Full Circle

The following are solutions to the practice problems presented earlier in this chapter.

1. Fill in the missing coordinates on the unit circle, represented by the letters. **A:** $\left(\frac{\sqrt{2}}{2}, \frac{\sqrt{2}}{2}\right)$; **B:** $\left(-\frac{1}{2}, \frac{\sqrt{3}}{2}\right)$; **C:** $\left(-\frac{\sqrt{3}}{2}, \frac{1}{2}\right)$; **D:** $\left(-\frac{\sqrt{2}}{2}, -\frac{\sqrt{2}}{2}\right)$; **E:** $\left(\frac{1}{2}, -\frac{\sqrt{3}}{2}\right)$; **F:** $\left(\frac{\sqrt{3}}{2}, -\frac{1}{2}\right)$.

2. Fill in the coordinates of the points on the unit circle that lie on the axes. They're represented by the letters A, B, C, and D. **A: (1, 0); B: (0, 1); C: (–1, 0); D: (0, –1).**

3. Find the y coordinate of the point on the unit circle if $x = \frac{12}{13}$, and the point lies in Quadrant I. $\left(\frac{\mathbf{12}}{\mathbf{13}}, \frac{\mathbf{5}}{\mathbf{13}}\right)$.

   Substitute this value into the equation for the unit circle, and then solve for y:

   $$\left(\frac{12}{13}\right)^2 + y^2 = 1$$
   $$\frac{144}{169} + y^2 = 1$$
   $$y^2 = 1 - \frac{144}{169} = \frac{25}{169}$$
   $$y = \pm\frac{5}{13}$$

   There are two answers, but the point is in Quadrant I, so use the positive answer. The point is $\left(\frac{12}{13}, \frac{5}{13}\right)$.

4. Find the x coordinate of the point on the unit circle if $y = \frac{7}{25}$, and the point lies in Quadrant II. $\left(-\frac{\mathbf{24}}{\mathbf{25}}, \frac{\mathbf{7}}{\mathbf{25}}\right)$.

   Substitute this value into the equation for the unit circle, and then solve for x:

   $$x^2 + \left(\frac{7}{25}\right)^2 = 1$$
   $$x^2 + \frac{49}{625} = 1$$
   $$x^2 = 1 - \frac{49}{625} = \frac{576}{625}$$
   $$x = \pm\frac{24}{25}$$

   There are two answers, but the point is in Quadrant II, so use the negative answer. The point is $\left(-\frac{24}{25}, \frac{7}{25}\right)$.

5. Find the y coordinate of the point on the unit circle if $x = -\frac{3}{5}$, and the point lies in Quadrant III. $\left(-\frac{\mathbf{3}}{\mathbf{5}}, -\frac{\mathbf{4}}{\mathbf{5}}\right)$.

   Substitute this value into the equation for the unit circle, and then solve for y:

   $$\left(-\frac{3}{5}\right)^2 + y^2 = 1$$
   $$\frac{9}{25} + y^2 = 1$$
   $$y^2 = 1 - \frac{9}{25} = \frac{16}{25}$$
   $$y = \pm\frac{4}{5}$$

   There are two answers, but the point is in Quadrant III, so use the negative answer. The point is $\left(-\frac{3}{5}, -\frac{4}{5}\right)$.

6 Find the x coordinate of the point on the unit circle if $y = -\frac{8}{17}$, and the point lies in Quadrant IV. $\left(\frac{15}{17}, -\frac{8}{17}\right)$.

Substitute this value into the equation for the unit circle, and then solve for x:

$$x^2 + \left(-\frac{8}{17}\right)^2 = 1$$
$$x^2 + \frac{64}{289} = 1$$
$$x^2 = 1 - \frac{64}{289} = \frac{225}{289}$$
$$x = \pm\frac{15}{17}$$

There are two answers, but the point is in Quadrant IV, so use the positive answer. The point is $\left(\frac{15}{17}, -\frac{8}{17}\right)$.

7 Find the y coordinate of the point on the unit circle if $x = \frac{1}{3}$, and the point lies in Quadrant I. $\left(\frac{1}{3}, \frac{2\sqrt{2}}{3}\right)$.

Substitute this value into the equation for the unit circle, and then solve for y:

$$\left(\frac{1}{3}\right)^2 + y^2 = 1$$
$$\frac{1}{9} + y^2 = 1$$
$$y^2 = 1 - \frac{1}{9} = \frac{8}{9}$$
$$y = \pm\frac{\sqrt{8}}{3} = \pm\frac{2\sqrt{2}}{3}$$

There are two answers, but the point is in Quadrant I, so use the positive answer. The point is $\left(\frac{1}{3}, \frac{2\sqrt{2}}{3}\right)$.

8 Find the x coordinate of the point on the unit circle if $y = -\frac{2}{7}$, and the point lies in Quadrant III. $\left(-\frac{3\sqrt{5}}{7}, -\frac{2}{7}\right)$.

Substitute this value into the equation for the unit circle, and then solve for x:

$$x^2 + \left(-\frac{2}{7}\right)^2 = 1$$
$$x^2 + \frac{4}{49} = 1$$
$$x^2 = 1 - \frac{4}{49} = \frac{45}{49}$$
$$x = \pm\frac{\sqrt{45}}{7} = \pm\frac{3\sqrt{5}}{7}$$

There are two answers, but the point is in Quadrant III, so use the negative answer. The point is $\left(-\frac{3\sqrt{5}}{7}, -\frac{2}{7}\right)$.

9 Find the reference angle for 120 degrees. **60 degrees.**

180 – 120 = 60 degrees.

10 Find the reference angle for 315 degrees. **45 degrees.**

360 – 315 = 45 degrees.

11 Find the reference angle for 210 degrees. **30 degrees.**

210 – 180 = 30 degrees.

12 Find the reference angle for $\frac{5\pi}{4}$. $\boldsymbol{\frac{\pi}{4}}$**.**

$\frac{5\pi}{4} - \pi = \frac{\pi}{4}$

13 Find the reference angle for $\frac{5\pi}{6}$. $\boldsymbol{\frac{\pi}{6}}$**.**

$\pi - \frac{5\pi}{6} = \frac{\pi}{6}$

14 Find the reference angle for $\frac{11\pi}{3}$. $\boldsymbol{\frac{\pi}{3}}$**.**

First, subtract two full rotations: $\frac{11\pi}{3} - 2\pi = \frac{11\pi}{3} - \frac{6\pi}{3} = \frac{5\pi}{3}$. Then $2\pi - \frac{5\pi}{3} = \frac{\pi}{3}$.

15 Which functions are positive for an angle measuring 105 degrees? **Sine and cosecant.**

16 Which functions are positive for an angle measuring 330 degrees? **Cosine and secant.**

17 Which functions are positive for an angle measuring $\frac{4\pi}{3}$? **Tangent and cotangent.**

18 Which functions are positive for an angle measuring $\frac{11\pi}{6}$? **Cosine and secant.**

19 Which functions are positive for an angle measuring $\frac{7\pi}{3}$? **Sine, cosine, tangent, cotangent, secant, cosecant.**

All the functions are positive for the angle — its terminal side is in Quadrant I.

20 Which functions are positive for an angle measuring 927 degrees? **Tangent and cotangent.**

The terminal side of this angle is in Quadrant III, so the tangent and cotangent are positive.

21 Find sin 315. $\boldsymbol{\sin 315 = -\frac{\sqrt{2}}{2}}$**.**

First select the quadrant for the terminal side. The terminal side is in Quadrant IV, and the sine is negative in that quadrant. The reference angle is 360 – 315 = 45 degrees. The sine of 45 degrees is $\frac{\sqrt{2}}{2}$, so $\sin 315 = -\frac{\sqrt{2}}{2}$.

22 Find cos 240. $\boldsymbol{-\frac{1}{2}}$**.**

First select the quadrant for the terminal side. The terminal side is in Quadrant III, and the cosine is negative in that quadrant. The reference angle is 240 – 180 = 60. The cosine of 60 degrees is $\frac{1}{2}$, so $\cos 240 = -\frac{1}{2}$.

23 Find tan $\frac{7\pi}{6}$. $\boldsymbol{\tan\frac{7\pi}{6} = \frac{\sqrt{3}}{3}}$**.**

First find the quadrant for the terminal side. The terminal side is in Quadrant III, and the tangent is positive in that quadrant. The reference angle is $\frac{7\pi}{6} - \pi = \frac{\pi}{6}$. The tangent of $\frac{\pi}{6}$ is $\frac{\sqrt{3}}{3}$, so $\tan\frac{7\pi}{6} = \frac{\sqrt{3}}{3}$.

24 Find csc $\frac{5\pi}{6}$. **2.**

The terminal side is in Quadrant II. The cosecant is positive in that quadrant. The reference angle is $\pi - \frac{5\pi}{6} = \frac{\pi}{6}$. The cosecant of $\frac{\pi}{6}$ is 2, so $\csc\frac{5\pi}{6} = 2$.

Part III
Trigonometric Identities and Equations

"This is my old trigonometry teacher, Mr. Wendt, his wife Doris, and their two children, Obtuse and Acute."

In this part . . .

I could've called this part AWOL: Acting With Obvious Levelheadedness. The trig identities offer so many opportunities for you. I show you how to change the identities and then apply them to simplify expressions and solve equations. I show you the different options and help you make good decisions. Making the correct choices and using the identities wisely are so important in trig.

Chapter 9

Identifying the Basic Identities

In This Chapter

- Reciprocating with the reciprocal identities
- Looking at ratios
- Working with Pythagorean identities

One thing that makes trig statements and equations so interesting is their flexibility and how many ways you can write them. If you don't like what you see, just replace it with something equivalent. There are several classifications of trigonometric identities — three of which are covered in this chapter.

Using the Reciprocal Identities

The reciprocal identities deal with the ways that the six basic trig functions can be paired up — three pairs of functions with their reciprocals (multiplicative inverses). The reciprocal identities are:

$$\sin x = \frac{1}{\csc x},\ \csc x = \frac{1}{\sin x}$$

$$\cos x = \frac{1}{\sec x},\ \sec x = \frac{1}{\cos x}$$

$$\tan x = \frac{1}{\cot x},\ \cot x = \frac{1}{\tan x}$$

You replace one function with another using identities in order to solve an equation or prepare to do some other mathematical process. This section provides practice doing the substitutions using identities.

Q. Use reciprocal identities to rewrite this in terms of just one trig function: $\frac{2\sin x}{\csc x}$.

A. Replace csc x with $\frac{1}{\sin x}$. This creates a complex fraction that can be simplified by multiplying the numerator by the reciprocal of the denominator (that's *reciprocal* or *flip* — not reciprocal function):

$$\frac{2\sin x}{\frac{1}{\sin x}} = \frac{2\sin x}{1} \cdot \frac{\sin x}{1} = 2\sin^2 x.$$

Notice the exponent of 2 between sin and x. That's a trig shorthand way of saying that the entire function sin x is being squared. $\sin^2 x = (\sin x)^2$. It saves on a lot of parentheses.

1. Use a reciprocal identity to rewrite the expression in terms of one function: $2\sin x + \frac{3}{\csc x}$.

2. Use a reciprocal identity to rewrite the expression in terms of one function: $\frac{1}{\cos x} + \frac{\sec x}{2}$.

Solve It

3. Use a reciprocal identity to rewrite the expression in terms of one function: $3\tan x - \frac{1}{3\cot x}$.

Solve It

4. Use a reciprocal identity to rewrite the expression in terms of one function: $\frac{\sin^2 x}{\csc^3 x}$.

Solve It

Creating the Ratio Identities

The ratio identities are the two identities in which the tangent and cotangent functions are defined in terms of sine and cosine. The two ratios, tangent and cotangent, are reciprocals of one another — just another confirmation of the relationship between the tangent and cotangent. The ratio identities are:

$$\tan x = \frac{\sin x}{\cos x}$$

$$\cot x = \frac{\cos x}{\sin x}$$

The ratio identities are helpful when solving identity problems where the technique being used is to change every function to sine and cosine.

Q. Simplify the statement using ratio identities: $\tan x\,(3\cos x) + \frac{2}{\cot x}(\cos x)$.

A. First, replace tan x and cot x with their ratio identities: $\frac{\sin x}{\cos x}(3\cos x) + \frac{2}{\frac{\cos x}{\sin x}}(\cos x)$. Simplify by multiplying the fractions in the two terms — rewriting the complex fraction on the right as a product with the reciprocal of the denominator:

$$\frac{\sin x}{\cancel{\cos x}} \cdot \frac{3\cancel{\cos x}}{1} + \frac{2}{1} \cdot \frac{\sin x}{\cancel{\cos x}} \cdot \frac{\cancel{\cos x}}{1} =$$

$3\sin x + 2\sin x = 5\sin x$.

5. Simplify the statement using ratio identities: $2\tan x\,(3\cos x)$.

6. Simplify the statement using ratio identities: $4\cot x + \frac{4}{\sin x}$.

Solve It

7. Simplify the statement using ratio identities: $5\cot x\left(\frac{3}{\cos x}\right)$.

Solve It

8. Simplify the statement using any identities: $\cot x\left(\frac{1}{3\cos x}+\frac{2}{\cot x}-\tan x\right)$.

Solve It

Playing Around with Pythagorean Identities

There are three Pythagorean identities. The first of these identities, the one involving sine and cosine, is frequently used in all sorts of work dealing with trigonometric statements. In the following listing of the identities, I've also included some altered or secondary versions of each of the identities. The Pythagorean identities are:

$$\sin^2 x + \cos^2 x = 1$$
$$\sin^2 x = 1 - \cos^2 x$$
$$\cos^2 x = 1 - \sin^2 x$$

$$\tan^2 x + 1 = \sec^2 x$$
$$\tan^2 x = \sec^2 x - 1$$

$$1 + \cot^2 x = \csc^2 x$$
$$\cot^2 x = \csc^2 x - 1$$

Q. Use a Pythagorean identity to simplify the statement: $\cos x\left(\tan^2 x\right)+\cos x$.

A. First, factor the cos x out of each term: $\cos x\left(\tan^2 x+1\right)$. Next, replace the $\tan^2 x+1$ using the appropriate Pythagorean identity: $\cos x\left(\sec^2 x\right)$. Recognizing that cosine and secant are reciprocals, rewrite the cosine factor using a reciprocal identity and simplify:

$$\frac{1}{\cancel{\sec x}}\cdot\frac{\sec^{\cancel{2}1} x}{1}=\sec x.$$

9. Simplify the expression using a Pythagorean identity: $(\sin x + \cos x)^2 - 1$.

Solve It

10. Simplify the expression using a Pythagorean identity: $\sin x\left(\sin x + \frac{\cos^2 x}{\sin x}\right)$.

Solve It

11. Simplify the expression using a Pythagorean identity: $\sin^4 x - \cos^4 x$. (***Hint:*** Factor as the difference between squares.)

Solve It

12. Simplify the expression using a Pythagorean identity: $\sin^2(\cot^2 x - \csc^2 x)$.

13. Simplify the expression using a Pythagorean identity: $\frac{\sin^2 x - 1}{\cos^2 x}$.

Solve It

14. Simplify the expression using a Pythagorean identity: $\frac{\sec^2 x - \tan^2 x}{\sin x}$.

Solve It

15. Simplify the expression using a Pythagorean identity: $3\cot^2 x\left(\tan^2 x + 1\right)$.

Solve It

16. Simplify the expression using a Pythagorean identity: $\frac{1}{\csc^2 x} + \frac{\cos x}{\sec x}$.

Solving Identities Using Reciprocals, Ratios, and Pythagoras

To *solve* an identity means to show that, in fact, it is a true statement — that one side of the equation is equal to the other side of the equation. Many techniques can be used, but the common thread through all of them is using the proper substitutions. You have to look at the statement and decide what would work best (that is, most quickly and efficiently). You won't always make the best choice — that comes with practice — but, if you always use a *correct* substitution and do *correct* manipulations, you won't destroy the truth of the statement. You just may not make it any better. Here are some examples using better choices.

Q. Show that $\sin^4 x - \cos^4 x = 2\sin^2 x - 1$ is an identity.

A. First, factor the two terms on the left as the difference of two squares. Then replace the first factor on the left with 1, using the Pythagorean identity. Then replace the $\cos^2 x$ using the Pythagorean identity; distribute and simplify. The two sides match:

$$\begin{aligned}
\sin^4 x - \cos^4 x &= 2\sin^2 x - 1\\
(\sin^2 x + \cos^2 x)(\sin^2 x - \cos^2 x) &=\\
(1)(\sin^2 x - \cos^2 x) &=\\
\sin^2 x - \cos^2 x &=\\
\sin^2 x - (1 - \sin^2 x) &=\\
\sin^2 x - 1 + \sin^2 x &=\\
2\sin^2 x - 1 &= 2\sin^2 x - 1
\end{aligned}$$

17. Show that $\sin x(\csc x - \sin x) = \cos^2 x$ is an identity.

Solve It

18. Show that $\cot^2 x - \csc^2 x = -1$ is an identity.

Solve It

19. Show that $(1 + \tan x)^2 - \sec^2 x = 2\tan x$ is an identity.

20. Show that $\frac{\csc x}{\cos x} - \frac{\cos x}{\sin x} = \tan x$ is an identity.

Solve It

21. Show that $\tan^2 x - \tan^2 x \sin^2 x = \sin^2 x$ is an identity.

Solve It

22. Show that $\tan^4 x + 2\tan^2 x + 1 = \sec^4 x$ is an identity.

Solve It

23. Show that $6\sec^2 x - 6\tan^2 x = 6$ is an identity.

Solve It

24. Show that $\frac{1 - \sin^2 x}{1 - \cos^2 x} = \cot^2 x$ is an identity.

Solve It

Answers to Problems on Basic Identities

The following are solutions to the practice problems presented earlier in this chapter.

1 Use a reciprocal identity to rewrite the expression in terms of one function: $2\sin x + \frac{3}{\csc x}$. **5 sin *x*.**

Replace the cosecant with its reciprocal. Then simplify the complex fraction by multiplying the numerator by the reciprocal of the denominator. Finish by adding the two like terms together:

$$\begin{aligned}2\sin x + \frac{3}{\frac{1}{\sin x}} &= 2\sin x + \frac{3}{1}\cdot\frac{\sin x}{1}\\ &= 2\sin x + 3\sin x\\ &= 5\sin x\end{aligned}$$

2 Use a reciprocal identity to rewrite the expression in terms of one function: $\frac{1}{\cos x} + \frac{\sec x}{2}$. $\mathbf{\frac{3}{2}}$ **sec *x*.**

Replace the cosine reciprocal with secant, and combine the terms:

$$\begin{aligned}\frac{1}{\cos x} + \frac{\sec x}{2} &= \sec x + \frac{1}{2}\sec x\\ &= \frac{3}{2}\sec x\end{aligned}$$

3 Use a reciprocal identity to rewrite the expression in terms of one function: $3\tan x - \frac{1}{3\cot x}$. $\mathbf{\frac{8}{3}}$ **tan *x*.**

Rewrite the cotangent first with the fraction in front; then replace the cotangent reciprocal with tangent. Combine the terms:

$$\begin{aligned}3\tan x - \frac{1}{3\cot x} &= 3\tan x - \frac{1}{3}\frac{1}{\cot x}\\ &= 3\tan x - \frac{1}{3}\tan x\\ &= \frac{8}{3}\tan x\end{aligned}$$

4 Use a reciprocal identity to rewrite the expression in terms of one function: $\frac{\sin^2 x}{\csc^3 x}$. **sin⁵ *x*.**

Rewrite the cosecant using its reciprocal. Then simplify the complex fraction:

$$\begin{aligned}\frac{\sin^2 x}{\csc^3 x} &= \frac{\sin^2 x}{\frac{1}{\sin^3 x}}\\ &= \frac{\sin^2 x}{1}\cdot\frac{\sin^3 x}{1}\\ &= \sin^5 x\end{aligned}$$

The shorthand notation for powers of trig functions is to put the power between the function abbreviation and the variable. This works for all powers except –1, because that would indicate an inverse function.

5 Simplify the statement using ratio identities: $2\tan x\,(3\cos x)$. **6 sin *x*.**

Rewrite the tangent function using sine divided by cosine, multiply, and simplify the product:

$$\begin{aligned}2\tan x\,(3\cos x) &= \frac{2}{1}\cdot\frac{\sin x}{\cancel{\cos x}}\cdot\frac{3}{1}\cdot\frac{\cancel{\cos x}}{1}\\ &= 6\sin x\end{aligned}$$

6 Simplify the statement using ratio identities: $4\cot x + \frac{4}{\sin x}$. $\mathbf{\frac{4\cos x + 4}{\sin x}}$.

Rewrite the cotangent using cosine divided by sine. Then add the two fractions that have a common denominator:

$$4\cot x+\frac{4}{\sin x}=\frac{4}{1}\cdot\frac{\cos x}{\sin x}+\frac{4}{\sin x}$$
$$=\frac{4\cos x}{\sin x}+\frac{4}{\sin x}$$
$$=\frac{4\cos x+4}{\sin x}$$

The numerator could be factored here, by factoring out a 4 from each term.

7 Simplify the statement using ratio identities: $5\cot x\left(\frac{3}{\cos x}\right)$. **15 csc *x*.**

Rewrite the cotangent using cosine divided by sine. Then simplify and multiply the fractions:

$$5\cot x\left(\frac{3}{\cos x}\right)=\frac{5}{1}\frac{\cancel{\cos x}}{\sin x}\left(\frac{3}{\cancel{\cos x}}\right)$$
$$=\frac{15}{\sin x}$$
$$=15\csc x$$

8 Simplify the statement using any identities: $\cot x\left(\frac{1}{3\cos x}+\frac{2}{\cot x}-\tan x\right)$. **⅓ csc *x* + 1.**

Distribute the cotangent function over the three terms in the parentheses. Then use the ratio identity for the first multiplication and a reciprocal identity for the last multiplication. Simplify to finish:

$$\cot x\left(\frac{1}{3\cos x}+\frac{2}{\cot x}-\tan x\right)=\cot x\left(\frac{1}{3\cos x}\right)+\cot x\left(\frac{2}{\cot x}\right)-\cot x(\tan x)$$
$$=\frac{\cancel{\cos x}}{\sin x}\left(\frac{1}{3\cancel{\cos x}}\right)+\frac{\cancel{\cot x}}{1}\left(\frac{2}{\cancel{\cot x}}\right)-\frac{\cancel{\cot x}}{1}\left(\frac{1}{\cancel{\cot x}}\right)$$
$$=\frac{1}{3\sin x}+2-1$$
$$=\frac{1}{3}\csc x+1$$

9 Simplify the expression using a Pythagorean identity: $(\sin x+\cos x)^2-1$. **2 sin *x* cos *x*.**

First square the binomial. Then pull out the two terms from the Pythagorean identity and substitute:

$$(\sin x+\cos x)^2-1=\sin^2 x+2\sin x\cos x+\cos^2 x-1$$
$$=(\sin^2 x+\cos^2 x)+2\sin x\cos x-1$$
$$=1+2\sin x\cos x-1$$
$$=2\sin x\cos x$$

10 Simplify the expression using a Pythagorean identity: $\sin x\left(\sin x+\frac{\cos^2 x}{\sin x}\right)$. **1.**

Distribute the sine over the two terms. Then use the Pythagorean identity to finish the simplification:

$$\sin x\left(\sin x+\frac{\cos^2 x}{\sin x}\right)=\sin x(\sin x)+\cancel{\sin x}\left(\frac{\cos^2 x}{\cancel{\sin x}}\right)$$
$$=\sin^2 x+\cos^2 x$$
$$=1$$

11 Simplify the expression using a Pythagorean identity: $\sin^4 x-\cos^4 x$. (***Hint:*** Factor as the difference between squares.) **(sin *x* – cos *x*)(sin *x* + cos *x*).**

After factoring, the Pythagorean theorem is used to simplify one of the factors. The resulting expression can also be factored, but there isn't another substitution to be done:

$$\begin{aligned}\sin^4 x - \cos^4 x &= (\sin^2 x - \cos^2 x)(\sin^2 x + \cos^2 x)\\ &= (\sin^2 x - \cos^2 x)(1)\\ &= (\sin x - \cos x)(\sin x + \cos x)\end{aligned}$$

12 Simplify the expression using a Pythagorean identity: $\sin^2 x(\cot^2 x - \csc^2 x)$. **$-\sin^2 x$.**

First distribute the sine factor over the other two. Use ratio and reciprocal identities to complete the multiplications:

$$\begin{aligned}\sin^2 x(\cot^2 x - \csc^2 x) &= \sin^2 x(\cot^2 x) - \sin^2 x(\csc^2 x)\\ &= \frac{\cancel{\sin^2 x}}{1}\left(\frac{\cos^2 x}{\cancel{\sin^2 x}}\right) - \frac{\cancel{\sin^2 x}}{1}\left(\frac{1}{\cancel{\sin^2 x}}\right)\\ &= \cos^2 x - 1\end{aligned}$$

Use the Pythagorean identity where you solve for the cosine term and substitute in:

$$\begin{aligned}\cos^2 x - 1 &= (1 - \sin^2 x) - 1\\ &= -\sin^2 x\end{aligned}$$

13 Simplify the expression using a Pythagorean identity: $\frac{\sin^2 x - 1}{\cos^2 x}$. **$-1$.**

Replace the sine term with its equivalent from the Pythagorean identity and simplify:

$$\begin{aligned}\frac{\sin^2 x - 1}{\cos^2 x} &= \frac{(1 - \cos^2 x) - 1}{\cos^2 x}\\ &= \frac{-\cos^2 x}{\cos^2 x}\\ &= -1\end{aligned}$$

 Simplify the expression using a Pythagorean identity: $\frac{\sec^2 x - \tan^2 x}{\sin x}$. **$\csc x$.**

Replace the secant term with its equivalent from the Pythagorean identity:

$$\begin{aligned}\frac{\sec^2 x - \tan^2 x}{\sin x} &= \frac{(\tan^2 x + 1) - \tan^2 x}{\sin x}\\ &= \frac{1}{\sin x}\\ &= \csc x\end{aligned}$$

15 Simplify the expression using a Pythagorean identity: $3\cot^2 x(\tan^2 x + 1)$. **$3\csc^2 x$.**

It might be tempting to distribute the cotangent here, but a better move is to use the Pythagorean identity to replace what's in the parentheses and then use ratio and reciprocal identities to complete the work:

$$\begin{aligned}3\cot^2 x(\tan^2 x + 1) &= 3\cot^2 x(\sec^2 x)\\ &= \frac{3}{1}\cdot\frac{\cancel{\cos^2 x}}{\sin^2 x}\cdot\frac{1}{\cancel{\cos^2 x}}\\ &= \frac{3}{\sin^2 x}\\ &= 3\csc^2 x\end{aligned}$$

16 Simplify the expression using a Pythagorean identity: $\frac{1}{\csc^2 x} + \frac{\cos x}{\sec x}$. **1.**

Use ratio identities, first, to simplify the terms. Then apply the Pythagorean identity to the result:

$$\begin{aligned}\frac{1}{\csc^2 x} + \frac{\cos x}{\sec x} &= \frac{1}{\frac{1}{\sin^2 x}} + \frac{\cos x}{\frac{1}{\cos x}}\\ &= \sin^2 x + \cos^2 x\\ &= 1\end{aligned}$$

 **17** Show that $\sin x(\csc x - \sin x) = \cos^2 x$ is an identity.

Distribute the sine term on the left. Then use a reciprocal identity to simplify the first term. A substitution using a Pythagorean identity finishes the job:

$$\begin{aligned}
\sin x(\csc x - \sin x) &= \cos^2 x\\
\sin x(\csc x) - \sin x(\sin x) &=\\
\sin x\left(\frac{1}{\sin x}\right) - \sin^2 x &=\\
1 - \sin^2 x &=\\
\cos^2 x &= \cos^2 x
\end{aligned}$$

18 Show that $\cot^2 x - \csc^2 x = -1$ is an identity.

Replace the cosecant term on the left with its equivalent in the Pythagorean identity. Distribute the negative and simplify:

$$\begin{aligned}
\cot^2 x - \csc^2 x &= -1\\
\cot^2 x - (1 + \cot^2 x) &=\\
\cot^2 x - 1 - \cot^2 x &=\\
-1 &= -1
\end{aligned}$$

19 Show that $(1 + \tan x)^2 - \sec^2 x = 2\tan x$ is an identity.

First square the binomial on the left. Then replace the secant term with its equivalent in the Pythagorean identity. Simplify what's left — you have two opposites that add to zero:

$$\begin{aligned}
(1 + \tan x)^2 - \sec^2 x &= 2\tan x\\
1 + 2\tan x + \tan x^2 - \sec^2 x &=\\
1 + 2\tan x + \tan^2 x - (\tan^2 x + 1) &=\\
1 + 2\tan x + \tan^2 x - \tan^2 x - 1 &=\\
2\tan x &= 2\tan x
\end{aligned}$$

20 Show that $\frac{\csc x}{\cos x} - \frac{\cos x}{\sin x} = \tan x$ is an identity.

First find a common denominator and change each fraction so it has that common denominator:

$$\begin{aligned}
\frac{\csc x}{\cos x} - \frac{\cos x}{\sin x} &= \tan x\\
\frac{\csc x}{\cos x}\cdot\frac{\sin x}{\sin x} - \frac{\cos x}{\sin x}\cdot\frac{\cos x}{\cos x} &=\\
\frac{\csc x \sin x}{\cos x \sin x} - \frac{\cos^2 x}{\cos x \sin x} &=
\end{aligned}$$

Simplify the numerator in the first fraction. Subtract the two fractions, writing the two terms in the same numerator. Then replace the numerator using the appropriate form of the Pythagorean identity and simplify:

$$\begin{aligned}
\frac{1}{\cos x \sin x} - \frac{\cos^2 x}{\cos x \sin x} &=\\
\frac{1 - \cos^2 x}{\cos x \sin x} &=\\
\frac{\sin^2 x}{\cos x \sin x} &=\\
\frac{\sin x}{\cos x} &=\\
\tan x &= \tan x
\end{aligned}$$

21 Show that $\tan^2 x - \tan^2 x \sin^2 x = \sin^2 x$ is an identity.

First, factor out the tangent from the two terms on the left. Replace the expression in the parentheses with its Pythagorean identity equivalent. Rewrite the tangent factor using the ratio identity and simplify:

$$\begin{aligned}\tan^2 x - \tan^2 x \sin^2 x &= \sin^2 x\\ \tan^2 x\left(1-\sin^2 x\right) &=\\ \tan^2 x\left(\cos^2 x\right) &=\\ \frac{\sin^2 x}{\cancel{\cos^2 x}}\cdot\frac{\cancel{\cos^2 x}}{1} &=\\ \sin^2 x &= \sin^2 x\end{aligned}$$

22 Show that $\tan^4 x + 2\tan^2 x + 1 = \sec^4 x$ is an identity.

Factor the expression on the left. It's a perfect square trinomial. Replace the expression in the parentheses with its Pythagorean identity equivalent, and square the result:

$$\begin{aligned}\tan^4 x + 2\tan^2 x + 1 &= \sec^4 x\\ \left(\tan^2 x + 1\right)^2 &=\\ \left(\sec^2 x\right)^2 &=\\ \sec^4 x &= \sec^4 x\end{aligned}$$

23 Show that $6\sec^2 x - 6\tan^2 x = 6$ is an identity.

First, factor out the 6. Then replace the secant term with its Pythagorean identity equivalent. Simplify the terms in the parentheses:

$$\begin{aligned}6\sec^2 x - 6\tan^2 x &= 6\\ 6\left(\sec^2 x - \tan^2 x\right) &=\\ 6\left(\tan^2 + 1 - \tan^2 x\right) &=\\ 6(1) &= 6\end{aligned}$$

24 Show that $\frac{1-\sin^2 x}{1-\cos^2 x} = \cot^2 x$ is an identity.

Replace both the numerator and the denominator with their equivalents, using the Pythagorean identity. Then apply the appropriate ratio identity:

$$\begin{aligned}\frac{1-\sin^2 x}{1-\cos^2 x} &= \cot^2 x\\ \frac{\cos^2 x}{\sin^2 x} &=\\ \cot^2 x &= \cot^2 x\end{aligned}$$

Another approach would be to replace just the denominator with its equivalent, rewrite the fraction as two terms, apply the reciprocal and ratio identities, and then use a Pythagorean identity. This takes longer, but works, too.

A goal in solving identities is to be as quick and efficient as possible.

Chapter 10

Using Identities Defined with Operations

In This Chapter

- Adding and subtracting in identities
- Multiplying two times or more in identities
- Taking half an angle in identities
- Solving identities with sums, differences, and multiple angles

The identities involving sums, differences, doubling, and other multiples all require you to determine the exact value of angles that are results of these operations on the more popularly used angles. These formulas are also valuable when changing the form of an expression to make it more workable in a particular situation. The groupings of identities have their similarities and differences.

Adding Up the Angles with Sum Identities

The identities for the function of the sum of two different angles allow you to write, for instance, an angle of 75 degrees as the sum of angles of 45 degrees and 30 degrees. The examples and problems here deal with angles you already have exact values for, but these identities can be applied to any angles whose functions are known or unknown.

The sum identities are:

$$\sin(A+B) = \sin A\cos B + \cos A\sin B$$

$$\cos(A+B) = \cos A\cos B - \sin A\sin B$$

$$\tan(A+B) = \frac{\tan A + \tan B}{1 - \tan A\tan B}$$

Q. Use the sum identity to show that sin 90 degrees is equal to sin (30 + 60) and sin (45 + 45).

A. From the charts and unit circle, sin 90 = 1. Using the sum identity with the sum of 30 + 60,

$$\sin(30+60) = \sin 30 \cos 60 + \cos 30 \sin 60$$
$$= \left(\frac{1}{2}\right)\left(\frac{1}{2}\right) + \left(\frac{\sqrt{3}}{2}\right)\left(\frac{\sqrt{3}}{2}\right)$$
$$= \frac{1}{4} + \frac{3}{4}$$
$$= 1$$

Then, using the sum identity with the sum of 45 + 45,

$$\sin(45+45) = \sin 45 \cos 45 + \cos 45 \sin 45$$
$$= \left(\frac{\sqrt{2}}{2}\right)\left(\frac{\sqrt{2}}{2}\right) + \left(\frac{\sqrt{2}}{2}\right)\left(\frac{\sqrt{2}}{2}\right)$$
$$= \frac{2}{4} + \frac{2}{4}$$
$$= 1$$

1. Use a sum identity to find sin 75 degrees.

Solve It

2. Use a sum identity to find tan 75 degrees.

Solve It

3. Use a sum identity to find $\cos \frac{7\pi}{12}$.

Solve It

4. Use a sum identity to find $\tan \frac{7\pi}{12}$.

Solve It

5. Use a sum identity to simplify $\sin(x + 90)$.

Solve It

6. Use a sum identity to simplify $\cos(x + 180)$.

Solve It

Subtracting Angles with Difference Identities

The difference identities work much like the sum identities. The actual identities will look familiar; they're pretty much the same as the sum identities, except that the signs are changed.

The angle difference identities:

$$\sin(A - B) = \sin A \cos B - \cos A \sin B$$

$$\cos(A - B) = \cos A \cos B + \sin A \sin B$$

$$\tan(A - B) = \frac{\tan A - \tan B}{1 + \tan A \tan B}$$

Some additional identities that go along with these difference identities are the opposite angle identities. I'll create two of them, here, by using the difference identity for sine and letting angle A be 0 degrees.

Q. Use the sine's angle difference identity to find sin (–*x*).

A.
$$\begin{aligned}\sin(-x) &= \sin(0 - x)\\ &= \sin 0 \cos x - \cos 0 \sin x\\ &= 0 \cdot \cos x - 1 \cdot \sin x\\ &= -\sin x\end{aligned}$$

Q. Use the cosine's angle difference identity to find cos (–*x*).

A.
$$\begin{aligned}\cos(-x) &= \cos(0 - x)\\ &= \cos 0 \cos x + \sin 0 \sin x\\ &= 1 \cdot \cos x + 0 \cdot \sin x\\ &= \cos x\end{aligned}$$

The sine of a negative angle is the opposite (negative) of the sine of the corresponding positive angle. With the cosine, though, the opposite angle and its corresponding positive angle have the same function value. The tangent works the same as the sine — it's a negative value.

7. Use an angle difference identity to find sin 15 degrees.

Solve It

8. Use an angle difference identity to find cos 15 degrees.

9. Use an angle difference identity to find $\tan \frac{\pi}{12}$.

Solve It

10. Use an angle difference identity to find $\sec \frac{\pi}{12}$.

Solve It

11. Simplify $\cos (90 - x)$.

Solve It

12. Simplify $\sin (180 - x)$.

Solve It

Doubling Your Pleasure with Double Angle Identities

The double angle identities allow you to find the function value of twice the angle — if you know the function value of the original angle. They're also used to simplify trigonometric expressions in physics and calculus to allow you to continue on to find the solution to a problem.

The double angle identities are:

$$\sin 2A = 2\sin A\cos A$$

$$\cos 2A = \cos^2 A - \sin^2 A$$

$$\tan 2A = \frac{2\tan A}{1-\tan^2 A}$$

The double angle formula for the cosine is also written in two other forms that are used even more frequently than this form. These forms can be created by substituting in for either the sine term or the cosine term in the Pythagorean identity $\sin^2 x + \cos^2 x = 1$, $\sin^2 x = 1 - \cos^2 x$, $\cos^2 x = 1 - \sin^2 x$.

Q. Rewrite the double angle formula for cosine by changing the functions to all sines.

A. $\cos 2A = \cos^2 A - \sin^2 A =$

$1 - \sin^2 x - \sin^2 x = 1 - 2\sin^2 x$

Q. Rewrite the double angle formula for cosine by changing the functions to all cosines.

A. $\cos 2A = \cos^2 A - \sin^2 A =$

$\cos^2 A - (1 - \cos^2 x) = 2\cos^2 x - 1$

13. Use a double angle identity to find sin 90 degrees.

Solve It

14. Use a double angle identity to find cos 540 degrees.

Solve It

15. Use a double angle identity to find cot 60 degrees.

Solve It

16. Use a double angle identity to find sec 90 degrees.

17. Demonstrate that $\tan 2x = \frac{\sin 2x}{\cos 2x}$.

Solve It

18. Use both a sum identity and double angle identity to demonstrate that $\sin(x+x) = \sin 2x$.

Multiplying the Many by Combining Sums and Doubles

The multiples of angles in identities aren't restricted to just twice an angle. You can also multiply an angle by 3, 4, 5, or whatever you want to find function values of those multiples of an angle. The sin 7*x* can be found by using sin (4*x* + 3*x*), and sin 4*x* can be found using sin 2(2*x*), while sin 3*x* = sin (2*x* + *x*). Whew! I'm not going to work through all the identities it would take to write one for sin 7*x*. I'd need sines and cosines of 4*x* and 3*x*, but the examples here show you the techniques needed.

Q. Use the sum identity and double identity for sine to find sin 3*x*.

A.
$$\begin{aligned}\sin 3x &= \sin(2x + x)\\ &= \sin 2x\cos x + \cos 2x \sin x\\ &= (2\sin x\cos x)\cos x + (1 - 2\sin^2 x)\sin x\\ &= 2\sin x\cos^2 x + \sin x - 2\sin^3 x\\ &= 2\sin x(1 - \sin^2 x) + \sin x - 2\sin^3 x\\ &= 2\sin x - 2\sin^3 x + \sin x - 2\sin^3 x\\ &= 3\sin x - 4\sin^3 x\end{aligned}$$

Notice how I've changed everything so they're in terms of sin *x*.

Q. Use the identity for 3*x* from the previous example to find sin 90, writing it as sin 3(30).

A. Keep in mind, here, that the *x* is the 30 for this application.

$$\begin{aligned}\sin 90 = \sin 3(30) &= 3\sin 30 - 4(\sin 30)^3\\ &= 3\left(\frac{1}{2}\right) - 4\left(\frac{1}{2}\right)^3\\ &= \frac{3}{2} - \frac{4}{8}\\ &= \frac{3}{2} - \frac{1}{2}\\ &= 1\end{aligned}$$

This is what we know the sine of 90 degrees to be.

Q. Use the double angle identity to find sin 4*x*.

A.
$$\begin{aligned}\sin 4x = \sin 2(2x) &= 2\sin 2x\cos 2x\\ &= 2(2\sin x\cos x)(1 - 2\sin^2 x)\\ &= 4\sin x\cos x - 8\sin^3 x\cos x\end{aligned}$$

This time, it wasn't as convenient to get the identity in terms of sine, only. I'll leave it like this rather than go into radical expressions.

Q. Use the identity for sin 4*x* to find sin 180, writing it as sin 4(45).

A.
$$\begin{aligned}\sin 180 = \sin 4(45) &= 4\sin 45\cos 45 - 8(\sin 45)^3\cos 45\\ &= 4\left(\frac{\sqrt{2}}{2}\right)\left(\frac{\sqrt{2}}{2}\right) - 8\left(\frac{\sqrt{2}}{2}\right)^3\left(\frac{\sqrt{2}}{2}\right)\\ &= 4\left(\frac{2}{4}\right) - 8\left(\frac{2\sqrt{2}}{8}\right)\left(\frac{\sqrt{2}}{2}\right)\\ &= 2 - 8\left(\frac{4}{16}\right)\\ &= 2 - 2\\ &= 0\end{aligned}$$

All that work for what you knew was going to come out to be 0. This just illustrates that you can use it for any angle — especially one you don't already know the answer to.

19. Use the sum identity $\sin(2x + x)$ and then the double-angle formula to find sin 90.

Solve It

20. Use the sum identity $\cos(2x + x)$ and then the double-angle formula to find cos 90.

Solve It

21. Find an identity for $\tan 4x$.

Solve It

22. Find an identity for $\cos 3x$.

Solve It

23. Find an identity for sin $(2x + 180)$.

Solve It

24. Find an identity for cos $(2x - 90)$.

Solve It

Halving Fun with Half-Angle Identities

The half-angle identities can be used to find function values for angles of 15 degrees (half of 30), 22.5 degrees (half of 45), and 7.5 degrees (half of 15 — which is half of 30). The half-angle identities are also used a lot in calculus to change from an expression that has a difficult solution to another form that's more manageable. Having these identities in your arsenal of possibilities is a good idea.

The half-angle identities are:

$$\sin \frac{x}{2} = \pm\sqrt{\frac{1-\cos x}{2}}$$

$$\cos \frac{\pi}{2} = \pm\sqrt{\frac{1+\cos x}{2}}$$

$$\tan \frac{x}{2} = \frac{1-\cos x}{\sin x} = \frac{\sin x}{1+\cos x}$$

The plus or minus part comes in to play when you determine which quadrant the terminal side lies in — you get to assign the sign.

Q. Find tan 30 degrees using a half-angle formula (half of 60).

A. It doesn't matter which version you use, so I'll just choose the first, because there's only one term in the denominator.

$$\tan 30 = \tan\frac{60}{2} = \frac{1-\cos 60}{\sin 60} = \frac{1-\frac{1}{2}}{\frac{\sqrt{3}}{2}} = \frac{\frac{1}{2}}{\frac{\sqrt{3}}{2}} = \frac{1}{2}\cdot\frac{2}{\sqrt{3}} = \frac{1}{\sqrt{3}} = \frac{\sqrt{3}}{3}$$

Q. Find cos 7.5 degrees using a half-angle formula twice.

A. $\cos 7.5 = \cos\frac{15}{2} = +\sqrt{\frac{1+\cos 15}{2}}$. I've chosen the + sign, because this angle is in Quadrant I. I'll have to find cos 15 to put into the formula.

$$\cos 15 = \cos\frac{30}{2} = +\sqrt{\frac{1+\cos 30}{2}} = \sqrt{\frac{1+\frac{\sqrt{3}}{2}}{2}} = \sqrt{\frac{2+\sqrt{3}}{4}} = \frac{\sqrt{2+\sqrt{3}}}{2}$$

Now, to finish this off,

$$\cos 7.5 = \sqrt{\frac{1+\frac{\sqrt{2+\sqrt{3}}}{2}}{2}} = \sqrt{\frac{2+\sqrt{2+\sqrt{3}}}{4}} = \frac{\sqrt{2+\sqrt{2+\sqrt{3}}}}{2}.$$

25. Use a half-angle formula to find sin 15 degrees.

Solve It

26. Use a half-angle formula to find cos 22.5 degrees.

27. Create a half-angle identity for cosecant.

28. Create an identity for $\sin \frac{x}{4}$.

Solve It

Simplifying Expressions with Identities

To simplify an expression in trigonometry means to make it simpler or more useable for the particular application. You may want to change it so only one function is used in the whole expression. You may want it factored so it can be set equal to zero or reduced in a fraction. Whatever the case, the trig identities are available to make the simplification possible.

Q. Simplify the expression $\sin^2 x \tan^2 x - \sin^2 x \sec^2 x$.

A. First, factor out the common sine factor in the two terms. Then replace the secant term with its equivalent in the Pythagorean identity.

$$\begin{aligned}\sin^2 x\left(\tan^2 x - \sec^2 x\right) &= \sin^2 x\left(\tan^2 x - \left[\tan^2 x + 1\right]\right)\\ &= \sin^2 x(-1)\\ &= -\sin^2 x\end{aligned}$$

This is nice, because only the one function is involved.

Q. Simplify the expression $\sin 2x - 2\cos x$.

A. There are two different angles involved, $2x$ and x. Use the double angle identity on the first term, and then factor out the common factors.

$$2\sin x\cos x - 2\cos x = 2\cos x(\sin x - 1)$$

This is far enough. There aren't any other convenient identities to use. The expression is factored, though, and ready for the next application.

29. Simplify the expression $2\cos^2 6x - 1$.

Solve It

30. Simplify the expression $\cos(30 - x) - \sin(60 - x)$.

31. Simplify the expression $\sin\left(\frac{\pi}{4} + x\right) - \cos\left(\frac{5\pi}{4} + x\right)$.

Solve It

32. Simplify the expression $\cos^2 2x + 4\sin^2 x\cos^2 x$.

Solving Identities

To *solve* an identity means to show that one side of the equation is equal to the other. These statements are true for all values of the variable that can be used with the functions involved. These are different from the equations in trigonometry where you have to find the specific angle measures that make the statement true. This process of solving identities involves doing substitutions using the basic identities. It also means using the rules of algebra and solving equations.

Q. Solve the identity $2\sin^2 x - \cos^2 x + 1 = 3\sin^2 x$.

A. I'm going to work on the left side, replacing the cosine term with its equivalence using a Pythagorean identity.

$$\begin{aligned} 2\sin^2 x - \cos^2 x + 1 &= 3\sin^2 x \\ 2\sin^2 x - (1 - \sin^2 x) + 1 &= \\ 2\sin^2 x - 1 + \sin^2 x + 1 &= \\ 3\sin^2 x &= 3\sin^2 x \end{aligned}$$

Q. Solve the identity $\frac{\cos x}{\sin x} + \frac{\sin x}{\cos x} = \sec x \csc x$.

A. I'll add the two fractions together after finding a common denominator and rewriting each fraction with that denominator.

$$\begin{aligned} \frac{\cos x}{\sin x} + \frac{\sin x}{\cos x} &= \sec x \csc x \\ \frac{\cos x}{\sin x} \cdot \frac{\cos x}{\cos x} + \frac{\sin x}{\cos x} \cdot \frac{\sin x}{\sin x} &= \\ \frac{\cos^2 x}{\sin x \cos x} + \frac{\sin^2 x}{\sin x \cos x} &= \end{aligned}$$

After adding the fractions, you find that the numerator is equal to 1, using the Pythagorean identity. Then just use the reciprocals to finish the job.

$$\begin{aligned} \frac{\cos^2 + \sin^2 x}{\sin x \cos x} &= \\ \frac{1}{\sin x \cos x} &= \\ \frac{1}{\sin x}\frac{1}{\cos x} &= \\ \csc x \sec x &= \sec x \csc x \end{aligned}$$

33. Solve the identity $\cos 3x = 4\cos^3 x - 3\cos x$.

Solve It

34. Solve the identity $\frac{\sin 2x}{1 + \cos 2x} = \tan x$.

Solve It

35. Solve the identity

$$\frac{\cos\frac{x}{2} + \sin\frac{x}{2}}{\cos\frac{x}{2} - \sin\frac{x}{2}} = \sec x + \tan x.$$

Solve It

36. Solve the identity $\cot 2x = \frac{\cot^2 x - 1}{2\cot x}$.

Solve It

37. Solve the identity
$$\frac{1}{\sec x + \tan x} = \sec x - \tan x.$$

Solve It

38. Solve the identity
$$\frac{\cos^2 x - \sin^2 x}{\sin x \cos x} = \cot x - \tan x.$$

Solve It

Answers to Problems on Using Identities

The following are solutions to the practice problems presented earlier in this chapter.

1 Use a sum identity to find sin 75 degrees. $\mathbf{\frac{\sqrt{6}+\sqrt{2}}{4}}$.

$$\begin{aligned}\sin(45+30) &= \sin 45 \cos 30 + \cos 45 \sin 30\\ &= \frac{\sqrt{2}}{2}\cdot\frac{\sqrt{3}}{2}+\frac{\sqrt{2}}{2}\cdot\frac{1}{2}\\ &= \frac{\sqrt{6}}{4}+\frac{\sqrt{2}}{4}=\frac{\sqrt{6}+\sqrt{2}}{4}\end{aligned}$$

This is an exact answer.

 Use a sum identity to find tan 75 degrees. $\mathbf{2+\sqrt{3}}$.

$$\begin{aligned}\tan(45+30) &= \frac{\tan 45+\tan 30}{1-\tan 45 \tan 30}\\ &= \frac{1+\frac{\sqrt{3}}{3}}{1-1\left(\frac{\sqrt{3}}{3}\right)}\\ &= \frac{\frac{3}{3}+\frac{\sqrt{3}}{3}}{\frac{3}{3}-\frac{\sqrt{3}}{3}}\\ &= \frac{3+\sqrt{3}}{3-\sqrt{3}}\end{aligned}$$

This is an exact answer, but it looks better if the denominator is rationalized:

$$\frac{3+\sqrt{3}}{3-\sqrt{3}}\cdot\frac{3+\sqrt{3}}{3+\sqrt{3}}=\frac{9+6\sqrt{3}+3}{9-3}=\frac{12+6\sqrt{3}}{6}=2+\sqrt{3}.$$

3 Use a sum identity to find $\cos\frac{7\pi}{12}$. $\mathbf{\frac{\sqrt{2}-\sqrt{6}}{4}}$.

$$\begin{aligned}\cos\left(\frac{\pi}{3}+\frac{\pi}{4}\right) &= \cos\frac{\pi}{3}\cos\frac{\pi}{4}-\sin\frac{\pi}{3}\sin\frac{\pi}{4}\\ &= \frac{1}{2}\cdot\frac{\sqrt{2}}{2}-\frac{\sqrt{3}}{2}\cdot\frac{\sqrt{2}}{2}\\ &= \frac{\sqrt{2}}{4}-\frac{\sqrt{6}}{4}\\ &= \frac{\sqrt{2}-\sqrt{6}}{4}\end{aligned}$$

4 Use a sum identity to find $\tan\frac{7\pi}{12}$. $\mathbf{-2-\sqrt{3}}$.

$$\begin{aligned}\tan\left(\frac{\pi}{3}+\frac{\pi}{4}\right) &= \frac{\tan\frac{\pi}{3}+\tan\frac{\pi}{4}}{1-\tan\frac{\pi}{3}\tan\frac{\pi}{4}}\\ &= \frac{\sqrt{3}+1}{1-\sqrt{3}\cdot 1}\\ &= \frac{1+\sqrt{3}}{1-\sqrt{3}}\end{aligned}$$

Rationalizing the denominator, $\frac{1+\sqrt{3}}{1-\sqrt{3}}\cdot\frac{1+\sqrt{3}}{1+\sqrt{3}}=\frac{1+2\sqrt{3}+3}{1-3}=\frac{4+2\sqrt{3}}{-2}=-2-\sqrt{3}.$

5 Use a sum identity to simplify sin $(x + 90)$. **cos *x*.**

$$\begin{aligned}\sin(x+90) &= \sin x\cos 90 + \cos x\sin 90\\ &= \sin x(0) + \cos x(1)\\ &= \cos x\end{aligned}$$

6 Use a sum identity to simplify cos $(x + 180)$. **–cos *x*.**

$$\begin{aligned}\cos(x+180) &= \cos x\cos 180 - \sin x\sin 180\\ &= \cos x(-1) - \sin x(0)\\ &= -\cos x\end{aligned}$$

7 Use an angle difference identity to find sin 15 degrees. $\mathbf{\frac{\sqrt{6}-\sqrt{2}}{4}}$.

$$\begin{aligned}\sin(45-30) &= \sin 45\cos 30 - \cos 45\sin 30\\ &= \frac{\sqrt{2}}{2}\cdot\frac{\sqrt{3}}{2} - \frac{\sqrt{2}}{2}\cdot\frac{1}{2}\\ &= \frac{\sqrt{6}}{4} - \frac{\sqrt{2}}{4}\\ &= \frac{\sqrt{6}-\sqrt{2}}{4}\end{aligned}$$

8 Use an angle difference identity to find cos 15 degrees. $\mathbf{\frac{\sqrt{6}+\sqrt{2}}{4}}$.

$$\begin{aligned}\cos(45-30) &= \cos 45\cos 30 + \sin 45\sin 30\\ &= \frac{\sqrt{2}}{2}\cdot\frac{\sqrt{3}}{2} + \frac{\sqrt{2}}{2}\cdot\frac{1}{2}\\ &= \frac{\sqrt{6}}{4} + \frac{\sqrt{2}}{4}\\ &= \frac{\sqrt{6}+\sqrt{2}}{4}\end{aligned}$$

9 Use an angle difference identity to find tan $\frac{\pi}{12}$. $\mathbf{2-\sqrt{3}}$.

$$\begin{aligned}\tan\left(\frac{\pi}{3}-\frac{\pi}{4}\right) &= \frac{\tan\frac{\pi}{3} - \tan\frac{\pi}{4}}{1+\tan\frac{\pi}{3}\tan\frac{\pi}{4}}\\ &= \frac{\sqrt{3}-1}{1+\sqrt{3}}\\ &= \frac{\sqrt{3}-1}{\sqrt{3}+1}\\ &= \frac{\sqrt{3}-1}{\sqrt{3}+1}\cdot\frac{\sqrt{3}-1}{\sqrt{3}-1}\\ &= \frac{3-2\sqrt{3}+1}{3-1}\\ &= \frac{4-2\sqrt{3}}{2}\\ &= 2-\sqrt{3}\end{aligned}$$

10 Use an angle difference identity to find sec $\frac{\pi}{12}$. $\mathbf{\sqrt{6}-\sqrt{2}}$.

The secant is the reciprocal of the cosine, so I'll find cos $\frac{\pi}{12}$, first, and then find the reciprocal of that answer.

$$\cos\left(\frac{\pi}{3}-\frac{\pi}{4}\right)=\cos\frac{\pi}{3}\cos\frac{\pi}{4}+\sin\frac{\pi}{3}\sin\frac{\pi}{4}$$
$$=\frac{1}{2}\cdot\frac{\sqrt{2}}{2}+\frac{\sqrt{3}}{2}\cdot\frac{\sqrt{2}}{2}$$
$$=\frac{\sqrt{2}}{4}+\frac{\sqrt{6}}{4}$$
$$=\frac{\sqrt{2}+\sqrt{6}}{4}$$

$$\sec\frac{\pi}{12}=\frac{1}{\cos\frac{\pi}{12}}=\frac{1}{\frac{\sqrt{2}+\sqrt{6}}{4}}=\frac{4}{\sqrt{2}+\sqrt{6}}$$

Rationalizing,

$$\frac{4}{\sqrt{2}+\sqrt{6}}\cdot\frac{\sqrt{2}-\sqrt{6}}{\sqrt{2}-\sqrt{6}}=\frac{4\left(\sqrt{2}-\sqrt{6}\right)}{2-6}$$
$$=\frac{4\left(\sqrt{2}-\sqrt{6}\right)}{-4}$$
$$=-\left(\sqrt{2}-\sqrt{6}\right)$$
$$=\sqrt{6}-\sqrt{2}$$

11 Simplify cos (90 – x). **sin *x*.**

$$\cos(90-x)=\cos 90\cos x+\sin 90\sin x$$
$$=(0)\cos x+(1)\sin x$$
$$=\sin x$$

12 Simplify sin (180 – x). **sin *x*.**

$$\sin(180-x)=\sin 180\cos x-\cos 180\sin x$$
$$=(0)\cos x-(-1)\sin x$$
$$=\sin x$$

13 Use a double angle identity to find sin 90 degrees. **1.**

$$\sin 90=\sin 2(45)=2\sin 45\cos 45$$
$$=2\left(\frac{\sqrt{2}}{2}\right)\left(\frac{\sqrt{2}}{2}\right)$$
$$=2\left(\frac{2}{4}\right)$$
$$=1$$

14 Use a double angle identity to find cos 540 degrees. **–1.**

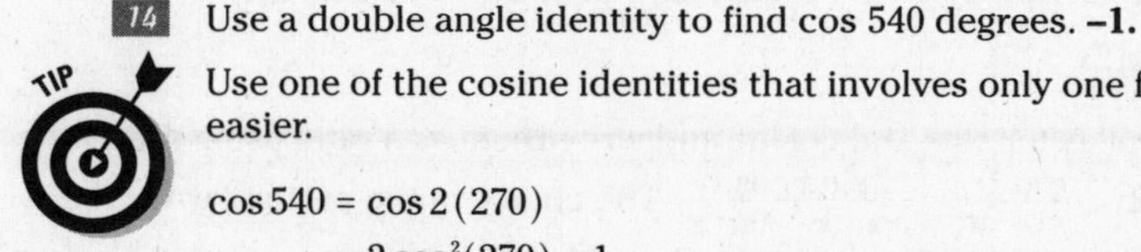

Use one of the cosine identities that involves only one function. It makes the computation easier.

$$\cos 540=\cos 2(270)$$
$$=2\cos^2(270)-1$$
$$=2(0)^2-1$$
$$=-1$$

The notation in trigonometry used to indicate the power of a function puts that power between the function name and the angle. So $\cos^2 x=(\cos x)^2$.

15 Use a double angle identity to find cot 60 degrees. $\frac{\sqrt{3}}{3}$**.**

The cotangent is the reciprocal of the tangent, so I'll use the double angle tangent identity and then find the reciprocal of the answer.

$$\begin{aligned}\tan 60 = \tan 2(30) &= \frac{2\tan 30}{1-\tan^2 30}\\ &= \frac{2\left(\frac{\sqrt{3}}{3}\right)}{1-\left(\frac{\sqrt{3}}{3}\right)^2}\\ &= \frac{\frac{2\sqrt{3}}{3}}{1-\frac{3}{9}}\\ &= \frac{\frac{2\sqrt{3}}{3}}{\frac{2}{3}}\\ &= \frac{2\sqrt{3}}{3}\cdot\frac{3}{2}\\ &= \sqrt{3}\end{aligned}$$

$$\cot 60 = \frac{1}{\tan 60} = \frac{1}{\sqrt{3}} = \frac{\sqrt{3}}{3}$$

16 Use a double angle identity to find sec 90 degrees. **Undefined.**

The secant is the reciprocal of the cosine, so find the cosine of 90 degrees, first, and then find the reciprocal.

$$\begin{aligned}\cos 90 = \cos 2(45) &= 2\cos^2 45 - 1\\ &= 2\left(\frac{\sqrt{2}}{2}\right)^2 - 1\\ &= 2\left(\frac{2}{4}\right) - 1\\ &= 1-1\\ &= 0\end{aligned}$$

$$\sec 90 = \frac{1}{\cos 90} = \frac{1}{0}$$

This has no value — you can't divide by 0. The secant isn't defined for 90 degrees.

17 Demonstrate that $\tan 2x = \frac{\sin 2x}{\cos 2x}$.

I'll use the double angle formulas for sine and cosine and write them as a ratio using the ratio identity for the tangent. $\tan 2x = \frac{\sin 2x}{\cos 2x} = \frac{2\sin x\cos x}{\cos^2 x - \sin^2 x}$. The choice of the particular version of the double angle identity for the cosine wasn't just random. I had in mind the next step — that I'll divide every term in the fraction by the square of cosine. (This is the same as multiplying both numerator and denominator by the reciprocal of the square.) Then I'll use the ratio identity to simplify the answer.

$$\frac{2\sin x\cos x}{\cos^2 x-\sin^2 x}=\frac{\frac{2\sin x\cos x}{\cos^2 x}}{\frac{\cos^2 x}{\cos^2 x}-\frac{\sin^2 x}{\cos^2 x}}$$
$$=\frac{\frac{2\sin x}{\cos x}\frac{\cos x}{\cos x}}{\frac{\cos^2 x}{\cos^2 x}-\frac{\sin^2 x}{\cos^2 x}}$$
$$=\frac{2\tan x}{1-\tan^2 x}$$

18 Use both a sum identity and double angle identity to demonstrate that $\sin(x+x)=\sin 2x$.

$\sin x\cos x+\cos x\sin x=2\sin x\cos x$

19 Use the sum identity sin $(2x + x)$ and then the double angle formula to find sin 90. **1.**

$$\sin 90=\sin\left[2(30)+30\right]$$
$$=\sin 2(30)\cos 30+\cos 2(30)\sin 30$$
$$=(2\sin 30\cos 30)\cos 30+(2\cos^2 30-1)\sin 30$$
$$=\left(2\cdot\frac{1}{2}\right)\cdot\frac{\sqrt{3}}{2}\cdot\frac{\sqrt{3}}{2}+\left[2\left(\frac{\sqrt{3}}{2}\right)^2-1\right]\cdot\frac{1}{2}$$
$$=\frac{\sqrt{3}\sqrt{3}}{4}+\left[\frac{2\cdot 3}{4}-1\right]\cdot\frac{1}{2}$$
$$=\frac{3}{4}+\frac{6-4}{4}\cdot\frac{1}{2}$$
$$=\frac{3}{4}+\frac{2}{4}\cdot\frac{1}{2}$$
$$=\frac{3}{4}+\frac{1}{4}$$
$$=1$$

Yes, I know that this is the hard way to find 1, but it illustrates how multiple angle identities work.

20 Use the sum identity cos $(2x + x)$ and then the double angle formula to find cos 90. **0.**

$$\cos 90=\cos\left[2(30)+30\right]$$
$$=\cos 2(30)\cos 30-\sin 2(30)\sin 30$$
$$=\left[2\cos^2 30-1\right]\cos 30-\left[2\sin 30\cos 30\right]\sin 30$$
$$=\left[2\left(\frac{\sqrt{3}}{2}\right)^2-1\right]\frac{\sqrt{3}}{2}-\left(2\cdot\frac{1}{2}\right)\cdot\frac{\sqrt{3}}{2}\cdot\frac{1}{2}$$
$$=\left[2\cdot\frac{3}{4}-1\right]\frac{\sqrt{3}}{2}-\frac{\sqrt{3}}{4}$$
$$=\left[\frac{3}{2}-1\right]\frac{\sqrt{3}}{2}-\frac{\sqrt{3}}{4}$$
$$=\frac{1}{2}\cdot\frac{\sqrt{3}}{2}-\frac{\sqrt{3}}{4}$$
$$=\frac{\sqrt{3}}{4}-\frac{\sqrt{3}}{4}$$
$$=0$$

21 Find an identity for tan $4x$. $\mathbf{\dfrac{4\tan x\left(1-\tan^2 x\right)}{1-6\tan^2 x+\tan^4 x}}$

Use tan 2(2x) and simplify it.

$$\begin{aligned}
\tan 2(2x) &= \frac{2\tan 2x}{1-\tan^2 2x}\\
&= \frac{2\left[\dfrac{2\tan x}{1-\tan^2 x}\right]}{1-\left[\dfrac{2\tan x}{1-\tan^2 x}\right]^2}\\
&= \frac{\dfrac{4\tan x}{1-\tan^2 x}}{1-\dfrac{4\tan^2 x}{\left(1-\tan^2 x\right)^2}}\\
&= \frac{4\tan x}{\cancel{1-\tan^2 x}}\cdot\frac{\left(1-\tan^2 x\right)^{\cancel{2}\,1}}{\left(1-\tan^2 x\right)^2-4\tan^2 x}\\
&= \frac{4\tan x\left(1-\tan^2 x\right)}{\left(1-2\tan^2 x+\tan^4 x\right)-4\tan^2 x}\\
&= \frac{4\tan x\left(1-\tan^2 x\right)}{1-6\tan^2 x+\tan^4 x}
\end{aligned}$$

22 Find an identity for cos $3x$. **$\cos x\,(1-4\sin^2 x)$.**

I'll use the sum identity with cos $(2x + x)$ and then make a careful choice when selecting the double angle identity for the cosine.

$$\begin{aligned}
\cos(2x+x) &= \cos 2x\cos x-\sin 2x\sin x\\
&= \left(1-2\sin^2 x\right)\cos x-2\sin x\cos x\sin x\\
&= \cos x-2\sin^2 x\cos x-2\sin^2 x\cos x\\
&= \cos x-4\sin^2 x\cos x\\
&= \cos x\left(1-4\sin^2 x\right)
\end{aligned}$$

It would have been much nicer to have this identity when doing problem 19. You can create identities for all sorts of multiples of angles, but you usually don't want to go to all the trouble unless you're going to have to use it many times.

23 Find an identity for sin $(2x + 180)$. **$-2\sin x\cos x$.**

$$\begin{aligned}
\sin(2x+180) &= \sin 2x\cos 180+\cos 2x\sin 180\\
&= \sin 2x(-1)+\cos 2x(0)\\
&= -\sin 2x\\
&= -2\sin x\cos x
\end{aligned}$$

I waited to apply the double-angle formula until I saw which term would still be there (not multiplied by 0).

24 Find an identity for cos $(2x - 90)$. **$2\sin x\cos x$.**

$$\begin{aligned}
\cos(2x-90) &= \cos 2x\cos 90+\sin 2x\sin 90\\
&= \cos 2x(0)+\sin 2x(1)\\
&= 0+\sin 2x\\
&= \sin 2x\\
&= 2\sin x\cos x
\end{aligned}$$

25 Use a half-angle formula to find sin 15 degrees. $\mathbf{\frac{\sqrt{2-\sqrt{3}}}{2}}$.

$$\begin{aligned}\sin 15 = \sin\frac{30}{2} &= +\sqrt{\frac{1-\cos 30}{2}}\\ &= \sqrt{\frac{1-\frac{\sqrt{3}}{2}}{2}}\\ &= \sqrt{\frac{2-\sqrt{3}}{4}}\\ &= \frac{\sqrt{2-\sqrt{3}}}{\sqrt{4}}\\ &= \frac{\sqrt{2-\sqrt{3}}}{2}\end{aligned}$$

I chose the + sign for the value, because the sine is positive in Quadrant I.

26 Use a half-angle formula to find cos 22.5 degrees. $\mathbf{\frac{\sqrt{2+\sqrt{2}}}{2}}$.

$$\begin{aligned}\cos 22.5 &= \cos\frac{45}{2}\\ &= +\sqrt{\frac{1+\cos 45}{2}}\\ &= \sqrt{\frac{1+\frac{\sqrt{2}}{2}}{2}}\\ &= \sqrt{\frac{2+\sqrt{2}}{4}}\\ &= \frac{\sqrt{2+\sqrt{2}}}{\sqrt{4}}\\ &= \frac{\sqrt{2+\sqrt{2}}}{2}\end{aligned}$$

The cosine is positive in Quadrant I.

27 Create a half-angle identity for cosecant. $\mathbf{\pm\sqrt{\frac{2}{1-\cos x}}}$

The cosecant is the reciprocal of the sine.

$$\csc\frac{x}{2} = \frac{1}{\sin\frac{x}{2}} = \frac{1}{\pm\sqrt{\frac{1-\cos x}{2}}} = \pm\sqrt{\frac{2}{1-\cos x}}$$

28 Create an identity for $\sin\frac{x}{4}$. $\mathbf{\pm\sqrt{\frac{\sqrt{2}\mp\sqrt{1+\cos x}}{2\sqrt{2}}}}$

$$\begin{aligned}\sin\frac{x}{4} &= \sin\frac{\frac{x}{2}}{2}\\ &= \pm\sqrt{\frac{1-\cos\frac{x}{2}}{2}}\\ &= \pm\sqrt{\frac{1\mp\sqrt{\frac{1+\cos x}{2}}}{2}}\\ &= \pm\sqrt{\frac{\sqrt{2}\mp\sqrt{1+\cos x}}{2\sqrt{2}}}\end{aligned}$$

This isn't too awfully helpful. It'd probably be just as easy to apply the identity twice in a problem rather than use this identity. Did you notice how the + or – became – or + when there was a – sign in front?

29 Simplify the expression $2\cos^2 6x - 1$. **$\cos 12x$.**

You might recognize this as being the cosine of a double-angle formula — the double angle is $12x$. That's the nicest way to simplify the expression — it makes it a single term. $2\cos^2 6x - 1 = \cos 2(6x) = \cos 12x$.

30 Simplify the expression $\cos(30 - x) - \sin(60 - x)$. **$\sin x$.**

Apply the two different difference identities and simplify the resulting expression.

$$\begin{aligned}&\cos(30 - x) - \sin(60 - x)\\&= \cos 30 \cos x + \sin 30 \sin x - (\sin 60 \cos x - \cos 60 \sin x)\\&= \frac{\sqrt{3}}{2}\cos x + \frac{1}{2}\sin x - \frac{\sqrt{3}}{2}\cos x + \frac{1}{2}\sin x\\&= \sin x\end{aligned}$$

31 Simplify the expression $\sin\left(\frac{\pi}{4} + x\right) - \cos\left(\frac{5\pi}{4} + x\right)$. **$\sqrt{2}\cos x$.**

Apply the two different sum identities and simplify the resulting expression.

$$\begin{aligned}&\sin\left(\frac{\pi}{4} + x\right) - \cos\left(\frac{5\pi}{4} + x\right)\\&= \sin\frac{\pi}{4}\cos x + \cos\frac{\pi}{4}\sin x - \left(\cos\frac{5\pi}{4}\cos x - \sin\frac{5\pi}{4}\sin x\right)\\&= \frac{\sqrt{2}}{2}\cos x + \frac{\sqrt{2}}{2}\sin x - \left(-\frac{\sqrt{2}}{2}\right)\cos x + \left(-\frac{\sqrt{2}}{2}\right)\sin x\\&= \sqrt{2}\cos x\end{aligned}$$

32 Simplify the expression $\cos^2 2x + 4\sin^2 x\cos^2 x$. **1.**

If you recognize the second term as being the square of a double-angle formula, then you can make short work of this. Otherwise, you can substitute in the double-angle formula of the first term, simplify, and get the same answer. I'll show you both ways. First,

$$\begin{aligned}\cos^2 2x + 4\sin^2 x\cos^2 x &= \cos^2 2x + (2\sin x\cos x)^2\\&= \cos^2 2x + \sin^2 2x\\&= 1\end{aligned}$$

Second,

$$\begin{aligned}\cos^2 2x + 4\sin^2 x\cos^2 x &= (\cos^2 x - \sin^2 x)^2 + 4\sin^2 x\cos^2 x\\&= \cos^4 x - 2\sin^2 x\cos^2 x + \sin^4 x + 4\sin^2 x\cos^2 x\\&= \cos^4 x + 2\sin^2 x\cos^2 x + \sin^4 x\\&= (\cos^2 x + \sin^2 x)^2\\&= 1^2\\&= 1\end{aligned}$$

33 Solve the identity $\cos 3x = 4\cos^3 x - 3\cos x$.

This looks very much like problem 22, but it isn't. The goal here is to have both sides be the same. Start with writing the left side as cos $(2x + x)$, but choose a different double-angle formula — you want everything in terms of cosine.

$$\begin{aligned}\cos(2x+x) &= 4\cos^3 x - 3\cos x\\ \cos 2x\cos x - \sin 2x\sin x &=\\ (2\cos^2 x - 1)\cos x - (2\sin x\cos x)\sin x &=\\ 2\cos^3 x - \cos x - 2\sin^2 x\cos x &=\\ 2\cos^3 x - \cos x - 2(1-\cos^2 x)\cos x &=\\ 2\cos^3 x - \cos x - 2\cos x + 2\cos^3 x &=\\ 4\cos^3 x - 3\cos x &= 4\cos^3 x - 3\cos x\end{aligned}$$

34 Solve the identity $\frac{\sin 2x}{1+\cos 2x} = \tan x$.

Apply the double-angle formulas and simplify the fraction on the left. You have to choose the double-angle formula for the cosine carefully. You want to get rid of that 1 in the denominator.

$$\begin{aligned}\frac{\sin 2x}{1+\cos 2x} &= \tan x\\ \frac{2\sin x\cos x}{1+2\cos^2 x - 1} &=\\ \frac{2\sin x\cos x}{2\cos^2 x} &=\\ \frac{\cancel{2}\sin x\cancel{\cos x}}{\cancel{2}\cos x\cancel{\cos x}} &=\\ \frac{\sin x}{\cos x} &=\\ \tan x &= \tan x\end{aligned}$$

35 Solve the identity $\frac{\cos\frac{x}{2} + \sin\frac{x}{2}}{\cos\frac{x}{2} - \sin\frac{x}{2}} = \sec x + \tan x$.

First, I'll insert the half-angle formulas.

$$\frac{\sqrt{\frac{1+\cos x}{2}} + \sqrt{\frac{1-\cos x}{2}}}{\sqrt{\frac{1+\cos x}{2}} - \sqrt{\frac{1-\cos x}{2}}} = \sec x + \tan x$$

The denominator of the fraction has two radicals in it. I'll rationalize the denominator.

$$\frac{\sqrt{\frac{1+\cos x}{2}} + \sqrt{\frac{1-\cos x}{2}}}{\sqrt{\frac{1+\cos x}{2}} - \sqrt{\frac{1-\cos x}{2}}} \cdot \frac{\sqrt{\frac{1+\cos x}{2}} + \sqrt{\frac{1-\cos x}{2}}}{\sqrt{\frac{1+\cos x}{2}} + \sqrt{\frac{1-\cos x}{2}}} = \sec x + \tan x$$

$$\frac{\frac{1+\cos x}{2} + 2\sqrt{\frac{1-\cos^2 x}{4}} + \frac{1-\cos x}{2}}{\frac{1+\cos x}{2} - \frac{1-\cos x}{2}} =$$

$$\frac{\frac{1+\cos x + 1 - \cos x}{2} + 2\sqrt{\frac{1-\cos^2 x}{2}}}{\frac{1+\cos x - 1 + \cos x}{2}} =$$

$$\frac{1+\sqrt{1-\cos^2 x}}{\cos x} =$$

Now, use the Pythagorean identity on the value under the radical, rewrite the fraction as two, and use the appropriate identities.

$$\frac{1+\sqrt{\sin^2 x}}{\cos x} =$$
$$\frac{1+\sin x}{\cos x} =$$
$$\frac{1}{\cos x} + \frac{\sin x}{\cos x} =$$
$$\sec x + \tan x = \sec x + \tan x$$

36 Solve the identity $\cot 2x = \frac{\cot^2 x - 1}{2\cot x}$.

The cotangent is the reciprocal of tangent, so I'll start with that reciprocal identity, insert the tangent double-angle formula, and change everything to cotangents.

$$\frac{1}{\tan 2x} = \frac{\cot^2 x - 1}{2\cot x}$$
$$\frac{1}{\frac{2\tan x}{1-\tan^2 x}} =$$
$$\frac{1-\tan^2 x}{2\tan x} =$$

One way to change to cotangents is to change all the tangent terms using the reciprocal identity. That will create a *complex fraction* (fractions within a fraction). Instead, a quicker, neater approach is to just multiply each term in the numerator and denominator by the square of $\cot x$. The product $(\tan x)(\cot x) = 1$.

$$\frac{1-\tan^2 x}{2\tan x} \cdot \frac{\cot^2 x}{\cot^2 x} =$$
$$\frac{1(\cot^2 x) - \tan^2 x(\cot^2 x)}{2\tan x(\cot^2 x)} =$$
$$\frac{\cot^2 x - 1}{2\cot x} = \frac{\cot^2 x - 1}{2\cot x}$$

37 Solve the identity $\frac{1}{\sec x + \tan x} = \sec x - \tan x$.

The expression on the right is the conjugate (same two terms, different sign between them) of the denominator. I'll multiply numerator and denominator of the fraction by that conjugate and then insert a Pythagorean identity.

$$\frac{1}{\sec x + \tan x} \cdot \frac{\sec x - \tan x}{\sec x - \tan x} = \sec x - \tan x$$
$$\frac{\sec x - \tan x}{\sec^2 x - \tan^2 x} =$$
$$\frac{\sec x - \tan x}{(\tan^2 x + 1) - \tan^2 x} =$$
$$\frac{\sec x - \tan x}{1} = \sec x - \tan x$$

38 Solve the identity $\frac{\cos^2 x - \sin^2 x}{\sin x \cos x} = \cot x - \tan x$.

Split up the fraction into two and reduce the fractions. The ratio identities will take care of the rest.

$$\frac{\cos^2 - \sin^2 x}{\sin x \cos x} = \cot x - \tan x$$
$$\frac{\cos^2 x}{\sin x \cos x} - \frac{\sin^2 x}{\sin x \cos x} =$$
$$\frac{\cos x \cancel{\cos x}}{\sin x \cancel{\cos x}} - \frac{\cancel{\sin x} \sin x}{\cancel{\sin x} \cos x} =$$
$$\frac{\cos x}{\sin x} - \frac{\sin x}{\cos x} =$$
$$\cot x - \tan x = \cot x - \tan x$$

Chapter 11

Techniques for Solving Trig Identities

In This Chapter

- Working on identities one side at a time
- Finding common denominators to solve identities
- Multiplying by conjugates
- Writing all functions in terms of just one

Solving trigonometric identities is good exercise for the mind. It's also a great skill to develop so you can become better at algebra, conquer trigonometry, and prepare for calculus. Take your pick, the reason is here.

Sometimes, the best method needed to solve a trig identity just leaps out at you. Other times, you just have to make an educated guess or jump in and try something. In this chapter, I discuss many different options that you can use to solve trig identities. Often, more than one technique will work on the same identity. Your goal is to pick the one that's quickest and easiest for you.

Working on One Side at a Time

The preferred method for solving identities is to work on just one side of the equation and make it match the other side. You look for the more complicated side to work on — the one with more terms or functions other than sine or cosine. It's usually easier to make two terms become one than the other way around. In the following two examples, I show you how to solve the same identity working from one side and then the other. You can decide which you think is easier and clearer.

Q. Solve the identity $\frac{\cot x + 4\sec x}{\cot x \csc x} = \sin x + 4\tan^2 x$, working on the left side.

A. First split up the fraction on the left into two, with the two different terms in the numerators and the same denominator. After that, reduce the first fraction and then change all the functions on the left using reciprocal identities:

$$\frac{\cot x + 4\sec x}{\cot x \csc x} = \sin x + 4\tan^2 x$$

$$\frac{\cot x}{\cot x \csc x} + \frac{4\sec x}{\cot x \csc x} =$$

$$\frac{{}^{1}\cancel{\cot x}}{{}_{1}\cancel{\cot x}\csc x} + \frac{4\sec x}{\cot x \csc x} =$$

$$\frac{1}{\frac{1}{\sin x}} + \frac{4\frac{1}{\cos x}}{\frac{\cos x}{\sin x}\frac{1}{\sin x}} =$$

Simplifying the complex fractions and then applying the ratio identity,

$$\sin x + \frac{4}{\cos x}\cdot\frac{\sin^2 x}{\cos x} =$$

$$\sin x + \frac{4\sin^2 x}{\cos^2 x} = \sin x + 4\tan^2 x$$

Q. Solve the same identity working on the right side.

A. First rewrite each term on the right as a fraction, find a common denominator, and add the fractions together.

$$\frac{\cot x + 4\sec x}{\cot x \csc x} = \sin x + 4\tan^2 x$$

$$= \frac{\sin x}{1} + \frac{4\sin^2 x}{\cos^2 x}$$

$$= \frac{\sin x \cos^2 x}{\cos^2 x} + \frac{4\sin^2 x}{\cos^2 x}$$

$$= \frac{\sin x \cos^2 x + 4\sin^2 x}{\cos^2 x}$$

Now divide each term in the fraction by the square of sine times cosine, simplify, and apply the appropriate ratio and reciprocal identities.

$$= \frac{\frac{\sin x \cos^2 x}{\sin^2 x \cos x} + \frac{4\sin^2 x}{\sin^2 x \cos x}}{\frac{\cos^2 x}{\sin^2 x \cos x}}$$

$$= \frac{\frac{\cos x}{\sin x} + \frac{4}{\cos x}}{\frac{\cos x}{\sin x}\frac{1}{\sin x}}$$

$$= \frac{\cot x + 4\sec x}{\cot x \csc x}$$

The hard part is coming up with the factor to divide everything by. It isn't always that obvious and can make you feel like a magician — pulling things out of thin air.

1. Solve the identity, working on one side only: $\frac{\tan x - 1}{1 - \cot x} = \tan x$.

Solve It

2. Solve the identity, working on one side only: $\frac{1 + \cot x}{\cot x} = \tan x + \csc^2 x - \cot^2 x$.

Solve It

3. Solve the identity, working on one side only: $\sin^2 2x + 2\cos 2x + 4\sin^4 x = 2$.

Solve It

4. Solve the identity, working on one side only: $\frac{1-\cos 8x}{2(1+\cos 4x)} = 1 - \cos 4x$.

Solve It

5. Solve the identity, working on one side only: $\frac{\sin x + \cos x}{\tan x} = \cos x + \frac{\cos^2 x}{\sin x}$.

Solve It

6. Solve the identity, working on one side only: $\frac{\sec^3 x + 3\sec x - 4\tan^2 x - 4}{\tan^2 x - 4\sec x + 4} = \sec x$.

Solve It

Working Back and Forth on Identities

The optimum situation when solving an identity is when you can work on just one side of the equation and get it to match the other side. That isn't always practical, though. Sometimes you have to work on each side separately to get them to be the same.

Q. Solve the identity $\frac{1-\cos x}{\csc x} = \frac{\sin^3 x}{1+\cos x}$.

A. On the left, rewrite the denominator using a reciprocal identity and then simplify the complex fraction. On the right, split up the cube of sine, use a Pythagorean identity, factor the difference of squares, and reduce.

$$\frac{1-\cos x}{\csc x} = \frac{\sin^3 x}{1+\cos x}$$

$$\frac{1-\cos x}{\frac{1}{\sin x}} = \frac{\sin x \sin^2 x}{1+\cos x}$$

$$\frac{1-\cos x}{1}\cdot\frac{\sin x}{1} = \frac{\sin x\left(1-\cos^2 x\right)}{1+\cos x}$$

$$(1-\cos x)\sin x = \frac{\sin x\left(1-\cos^2 x\right)}{1+\cos x}$$

$$= \frac{\sin x(1-\cos x)\cancel{(1+\cos x)}^{1}}{\cancel{1+\cos x}_{1}}$$

$$= \sin x(1-\cos x)$$

Q. Solve the identity $\tan(x+\pi) = \frac{\sec^2 x - 1}{\tan x}$.

A. This identity particularly begs to be solved by working on both sides at the same time. There don't seem to be any connections, otherwise. Use the tangent sum identity on the left, and use a Pythagorean identity in the numerator, on the right. The tangent of π is replaced with its numerical value, 0.

$$\tan(x+\pi) = \frac{\sec^2 x - 1}{\tan x}$$

$$\frac{\tan x + \tan\pi}{1-\tan x\tan\pi} = \frac{\left(\tan^2 x + 1\right) - 1}{\tan x}$$

$$\frac{\tan x + 0}{1-\tan x(0)} = \frac{\tan^2 x}{\tan x}$$

$$\frac{\tan x}{1} = \tan x$$

7. Solve the identity by working on each side at the same time: $(\sin x + \cos x)\sin x = 1 - \cos x(\cos x - \sin x)$.

Solve It

8. Solve the identity by working on each side at the same time: $\frac{\sin 2x + \cos 2x}{\sin x \cos x} = 2 + \cot x - \tan x$.

Solve It

9. Solve the identity by working on each side at the same time:

$$(\sin x + \cot x)^2 = \frac{\sin^4 x + \cos^2 x}{\sin^2 x} + 2\cos x.$$

Solve It

10. Solve the identity by working on each side at the same time:

$$\frac{\sin(x - \pi)}{\sin(x + \pi)} = 3 - 2\sin^2 x - 2\cos^2 x.$$

Solve It

Changing Everything to Sine and Cosine

The two most basic trigonometric functions are the sine and cosine. They're in the ratios for tangent and cotangent. They're the reciprocals of cosecant and secant. For this reason, a good technique to use when solving identities is to change every function until it only has sines and cosines in it. You have fewer variables to look at when solving the identity.

Q. Change everything to sines and cosines when solving the identity $\csc x - \cot x \cos x = \sin x$.

A. The two terms on the left will have a common denominator, so they can be subtracted. Then apply the Pythagorean identity.

$$\csc x - \cot x \cos x = \sin x$$

$$\frac{1}{\sin x} - \frac{\cos x}{\sin x}\frac{\cos x}{1} =$$

$$\frac{1 - \cos^2 x}{\sin x} =$$

$$\frac{\sin^2 x}{\sin x} =$$

$$\sin x = \sin x$$

Q. Change everything to sines and cosines when solving the identity $\frac{\sec^2 x}{\sec^2 x - 1} = \csc^2 x$.

A. After changing the functions, simplify the complex fraction on the left, and apply the Pythagorean identity.

$$\frac{\sec^2 x}{\sec^2 x - 1} = \csc^2 x$$

$$\frac{\frac{1}{\cos^2 x}}{\frac{1}{\cos^2 x} - 1} = \frac{1}{\sin^2 x}$$

$$\frac{\frac{1}{\cos^2 x}}{\frac{1 - \cos^2 x}{\cos^2 x}} =$$

$$\frac{1}{\cancel{\cos^2 x}_1} \cdot \frac{\cancel{\cos^2 x}^1}{1 - \cos^2 x} =$$

$$\frac{1}{\sin^2 x} = \frac{1}{\sin^2 x}$$

11. Solve the identity by changing all the functions to sines and cosines: $\sec^2 x + \csc^2 x = \sec^2 x \csc^2 x$.

Solve It

12. Solve the identity by changing all the functions to sines and cosines: $\dfrac{\cos x + \cot x}{\tan x + \sec x} = \cos x \cot x$.

13. Solve the identity by changing all the functions to sines and cosines: $\sec^2 x + \cot^2 x = \csc^2 x + \tan^2 x$.

Solve It

14. Solve the identity by changing all the functions to sines and cosines: $\dfrac{1 + \cot x}{\cot x} = \tan x + \csc^2 x - \cot^2 x$.

Multiplying by Conjugates

The technique of multiplying by conjugates comes up in several situations in mathematics. You use it with radicals, complex numbers, and now, in trigonometry. A *conjugate* of an expression is similar to the original expression. It has the same two variables or numbers, but the opposite sign is between them. The conjugate of $x + 4$ is $x - 4$. You multiply by conjugates when solving identities and you're stuck with a fraction with two terms in the denominator — and there's nothing else you can do with it.

Q. Solve the identity $\frac{1}{\csc x + \cot x} = \csc x - \cot x$.

A. Multiply the numerator and denominator of the fraction by the conjugate of the denominator. Then there is a Pythagorean identity that can be inserted.

$$\frac{1}{\csc x + \cot x} \cdot \frac{\csc x - \cot x}{\csc x - \cot x} = \csc x - \cot x$$

$$\frac{\csc x - \cot x}{\csc^2 x - \cot^2 x} =$$

$$\frac{\csc x - \cot x}{(1 + \cot^2 x) - \cot^2 x} =$$

$$\frac{\csc x - \cot x}{1} = \csc x - \cot x$$

Q. Solve the identity $\frac{1}{1 + \sec x} = \frac{\sec x - 1}{\tan^2 x}$.

A. Multiply the numerator and denominator of the fraction on the left by the conjugate of the denominator. Then apply the Pythagorean identity. Multiplying numerator and denominator of the result by –1 gives the terms the correct signs.

$$\frac{1}{1 + \sec x} \cdot \frac{1 - \sec x}{1 - \sec x} = \frac{\sec x - 1}{\tan^2 x}$$

$$\frac{1 - \sec x}{1 - \sec^2 x} =$$

$$\frac{1 - \sec x}{1 - (1 + \tan^2 x)} =$$

$$\frac{1 - \sec x}{-\tan^2 x} =$$

$$\frac{\sec x - 1}{\tan^2 x} = \frac{\sec x - 1}{\tan^2 x}$$

15. Solve the identity by multiplying by a conjugate: $\frac{\cot x}{\csc x - 1} = \frac{\csc x + 1}{\cot x}$.

Solve It

16. Solve the identity by multiplying by a conjugate: $\frac{\sin x}{1 + \cos x} = \csc x - \cot x$.

Solve It

17. Solve the identity by multiplying by a conjugate: $\frac{1-\sin x}{\cos x} = \frac{\cos x}{1+\sin x}$.

Solve It

18. Solve the identity by multiplying by a conjugate: $\frac{1+\sin x+\cos x}{1+\sin x-\cos x} = \frac{1+\cos x}{\sin x}$.

Solve It

Squaring Both Sides

Another version of working on both sides of an identity is squaring both sides. This works especially well when there's a radical in the equation that you need to get rid of.

Q. Solve the identity $2\sin^2 x - 1 = \sqrt{1-\sin^2 2x}$.

A. Square both sides to get rid of the radical. Then square the binomial on the left and apply the double-angle formula on the right.

$$(2\sin^2 x - 1)^2 = \left(\sqrt{1-\sin^2 2x}\right)^2$$

$$\begin{aligned}4\sin^4 x - 4\sin^2 x + 1 &= 1 - (\sin 2x)^2\\ &= 1 - (2\sin x\cos x)^2\\ &= 1 - 4\sin^2 x\cos^2 x\end{aligned}$$

Next, apply the Pythagorean identity.

$$\begin{aligned}&= 1 - 4\sin^2 x(1-\sin^2 x)\\ 4\sin^4 x - 4\sin^2 x + 1 &= 1 - 4\sin^2 x + 4\sin^4 x\end{aligned}$$

Q. Solve the identity $\sqrt{2(1+\cos 2x)} = \frac{\sin 2x}{\sin x}$.

A. Square both sides. Then apply the double angle identities.

$$\left(\sqrt{2(1+\cos 2x)}\right)^2 = \left(\frac{\sin 2x}{\sin x}\right)^2$$

$$2(1+2\cos^2 x - 1) = \left(\frac{2\cancel{\sin x}\cos x}{\cancel{\sin x}}\right)^2$$

$$2(2\cos^2 x) = (2\cos x)^2$$

$$4\cos^2 x = 4\cos^2 x$$

19. Solve the identity: $\sqrt{\frac{1-\sin x}{1+\sin x}} = \frac{\cos x}{1+\sin x}$.

Solve It

20. Solve the identity: $\sqrt{\frac{1+\cos x}{1-\cos x}} = \csc x + \cot x$.

Solve It

Finding Common Denominators

Before you can add or subtract fractions, you have to be sure that the denominators are the same. If they aren't, you determine their common denominator and rewrite the fractions. This process is actually very helpful when solving identities. It allows you to add those fractions together, and it's usually a big help in solving the identity. You use this method when you have two or more terms on one side and only one term on the other side of the equation. It's a way of putting the terms together.

Q. Solve the identity $\frac{1-\cos x}{\sin x} + \frac{\sin x}{1-\cos x} = 2\csc x$.

A. After rewriting the two fractions on the left with the common denominator and adding them together, square the binomial and use the Pythagorean identity. Simplify the numerator, reduce, and use the ratio identity.

$$\frac{1-\cos x}{\sin x}\cdot\frac{1-\cos x}{1-\cos x} + \frac{\sin x}{1-\cos x}\cdot\frac{\sin x}{\sin x} = 2\csc x$$

$$\frac{(1-\cos x)^2 + \sin^2}{\sin x(1-\cos x)} =$$

$$\frac{1-2\cos x+\cos^2 x+1-\cos^2 x}{\sin x(1-\cos x)} =$$

$$\frac{2-2\cos x}{\sin x(1-\cos x)} =$$

$$\frac{2\cancel{(1-\cos x)}^1}{\sin x\cancel{(1-\cos x)}_1} =$$

$$\frac{2}{\sin x} =$$

$$2\csc x = 2\csc x$$

Q. Solve the identity $\frac{1+\sec x}{\tan x} - \frac{\tan x}{\sec x} = \frac{1+\sec x}{\sec x\tan x}$.

A. After rewriting the two fractions on the left with the common denominator and adding them together, simplify the numerator. Apply the Pythagorean identity and simplify again.

$$\frac{1+\sec x}{\tan x}\cdot\frac{\sec x}{\sec x} - \frac{\tan x}{\sec x}\cdot\frac{\tan x}{\tan x} = \frac{1+\sec x}{\sec x\tan x}$$

$$\frac{(1+\sec x)\sec x - \tan^2 x}{\tan x\sec x} =$$

$$\frac{\sec x+\sec^2 x-\tan^2 x}{\tan x\sec x} =$$

$$\frac{\sec x+(\tan^2 x+1)-\tan^2 x}{\tan x\sec x} =$$

$$\frac{\sec x+1}{\tan x\sec x} = \frac{1+\sec x}{\sec x\tan x}$$

21. Solve the identity:

$$\frac{1}{\sin x+1}-\frac{1}{\sin x-1}=2\sec^2 x.$$

Solve It

22. Solve the identity: $\frac{\sec x}{\sin x}+\frac{\csc x}{\cos x}=\frac{4}{\sin 2x}$.

Solve It

23. Solve the identity:

$$\frac{1}{\sec x-1}-\frac{1}{\sec x+1}=2\cot^2 x.$$

Solve It

24. Solve the identity:

$$\frac{\sin x}{1+\cos x}+\frac{1+\cos x}{\sin x}=2\csc x.$$

Solve It

Writing All Functions in Terms of Just One

Each of the six trig functions has an equivalence or identity in terms of each of the other trig functions. The sine function can be written as something equal to cosine or tangent or cotangent or secant or cosecant. Here's what they look like:

$$\sin x = \sqrt{1-\cos^2 x} \qquad \sin x = \frac{\tan x}{\sqrt{\tan^2 x + 1}}$$

$$\sin x = \frac{1}{\sqrt{1+\cot^2 x}} \qquad \sin x = \frac{\sqrt{\sec^2 x - 1}}{\sec x}$$

$$\sin x = \frac{1}{\csc x}$$

Writing a trig expression with just one function in it can simplify your work if you're having to put in function values for a lot of problems. You can just find the value for the one function and use it over and over. I gave you the equivalents for sine. Here's how I got a couple of them. In each case, I chose an identity that involved the sine and also the other function it's to be written in terms of. Sometimes you have to also use a reciprocal of the function you want.

Q. Write the sine in terms of cosine.

A. The sine and cosine are both in the Pythagorean identity.

$$\sin^2 x + \cos^2 x = 1$$
$$\sin^2 x = 1 - \cos^2 x$$

To solve for the sine, just take the square root of each side. The + or – value will be assigned when you know which quadrant the angle is in that you're working with. $\sin x = \sqrt{1-\cos^2 x}$.

Q. Write the sine in terms of the tangent.

A. The sine and tangent aren't alone in any standard identity. Start with the ratio identity involving the tangent and solve for the sine.

$$\tan x = \frac{\sin x}{\cos x}$$
$$\sin x = \cos x \tan x$$

Use the reciprocal identity on the cosine, so the expression is in terms of the tangent and secant. Then you can use the Pythagorean identity involving the tangent and secant to solve for the secant in terms of tangent and substitute it in.

$$\sin x = \frac{1}{\sec x}\tan x = \frac{\tan x}{\sec x}$$
$$\tan^2 x + 1 = \sec^2 x, \text{ so } \sec x = \sqrt{\tan^2 x + 1}$$
$$= \frac{\tan x}{\sqrt{\tan^2 x + 1}}$$

25. Write the cosine function in terms of sine.

Solve It

26. Write the cosine function in terms of tangent.

Solve It

27. Write the tangent function in terms of sine.

Solve It

28. Write the secant function in terms of cosecant.

Solve It

Answers to Problems Techniques for Solving Identities

The following are solutions to the practice problems presented earlier in this chapter.

1 Solve the identity, working on one side only: $\frac{\tan x - 1}{1 - \cot x} = \tan x$.

Rewrite the cotangent in the denominator using the reciprocal identity. Then multiply both numerator and denominator by tan *x*. Don't distribute it in the numerator — leave it factored.

$$\frac{\tan x - 1}{1 - \cot x} = \tan x$$

$$\frac{\tan x - 1}{1 - \frac{1}{\tan x}} =$$

$$\frac{\tan x(\tan x - 1)}{\tan x\left(1 - \frac{1}{\tan x}\right)} =$$

$$\frac{\tan x(\tan x - 1)}{\tan x - 1} =$$

Now just reduce the fraction.

$$\frac{\tan x\,\cancel{(\tan x - 1)}^{1}}{{}_{1}\cancel{\tan x - 1}} =$$

$$\tan x = \tan x$$

2 Solve the identity, working on one side only: $\frac{1 + \cot x}{\cot x} = \tan x + \csc^2 x - \cot^2 x$.

The right side has three different functions. Use a Pythagorean identity to rewrite the cosecant, and then simplify.

$$\begin{aligned}\frac{1 + \cot x}{\cot x} &= \tan x + \csc^2 x - \cot^2 x \\ &= \tan x + (1 + \cot^2 x) - \cot^2 x \\ &= \tan x + 1\end{aligned}$$

Now just rewrite the tangent in terms of its reciprocal, find a common denominator for the two terms, and add them together.

$$\begin{aligned}&= \frac{1}{\cot x} + 1 \\ &= \frac{1}{\cot x} + \frac{\cot x}{\cot x} \\ \frac{1 + \cot x}{\cot x} &= \frac{1 + \cot x}{\cot x}\end{aligned}$$

3 Solve the identity, working on one side only: $\sin^2 2x + 2\cos 2x + 4\sin^4 x = 2$.

Rewrite the double angle identities to get all the terms using the same angle.

$$\sin^2 2x + 2\cos 2x + 4\sin^4 x = 2$$

$$(2\sin x\cos x)^2 + 2(1 - 2\sin^2 x) + 4\sin^4 x =$$

$$4\sin^2 x\cos^2 x + 2 - 4\sin^2 x + 4\sin^4 x =$$

Next, replace the cosine term using the Pythagorean identity. Simplify. You'll notice that there are two pairs of terms and their opposites, which will add up to 0.

$$4\sin^2 x(1-\sin^2 x)+2-4\sin^2 x+4\sin^4 x=$$
$$4\sin^2 x-4\sin^4 x+2-4\sin^2 x+4\sin^4 x=$$
$$2=2$$

 Solve the identity, working on one side only: $\frac{1-\cos 8x}{2(1+\cos 4x)}=1-\cos 4x$.

Apply the double angle identity to cos 8*x*. Then simplify the numerator and reduce the fraction.

$$\frac{1-(2\cos^2 4x-1)}{2(1+\cos 4x)}=1-\cos 4x$$
$$\frac{1-2\cos^2 4x+1}{2(1+\cos 4x)}=$$
$$\frac{2-2\cos^2 4x}{2(1+\cos 4x)}=$$
$$\frac{\cancel{2}(1-\cos^2 4x)}{\cancel{2}(1+\cos 4x)}=$$

The numerator is the difference of two squares. Factor it and reduce the fraction.

$$\frac{(1-\cos 4x)\cancelto{1}{(1+\cos 4x)}}{\cancelto{1}{1+\cos 4x}}=$$
$$1-\cos 4x=1-\cos 4x$$

5 Solve the identity, working on one side only: $\frac{\sin x+\cos x}{\tan x}=\cos x+\frac{\cos^2 x}{\sin x}$.

Find a common denominator for the two terms on the right. Rewrite the fractions, and add them together. Then factor out cosine from the two terms in the numerator.

$$\frac{\sin x+\cos x}{\tan x}=\cos x+\frac{\cos^2 x}{\sin x}$$
$$=\frac{\cos x}{1}+\frac{\cos^2 x}{\sin x}$$
$$=\frac{\cos x}{1}\cdot\frac{\sin x}{\sin x}+\frac{\cos^2 x}{\sin x}$$
$$=\frac{\sin x\cos x}{\sin x}+\frac{\cos^2 x}{\sin x}$$
$$=\frac{\sin x\cos x+\cos^2 x}{\sin x}$$
$$=\frac{\cos x(\sin x+\cos x)}{\sin x}$$

Now multiply both numerator and denominator by $\frac{1}{\cos x}$. Simplify, and use the ratio identity on the denominator.

$$=\frac{\frac{1}{\cos x}}{\frac{1}{\cos x}}\cdot\frac{\cos x(\sin x+\cos x)}{\sin x}$$
$$=\frac{\frac{\cos x}{\cos x}(\sin x+\cos x)}{\frac{\sin x}{\cos x}}$$
$$\frac{\sin x+\cos x}{\tan x}=\frac{\sin x+\cos x}{\tan x}$$

6 Solve the identity, working on one side only: $\frac{\sec^3 x + 3\sec x - 4\tan^2 x - 4}{\tan^2 x - 4\sec x + 4} = \sec x$.

Change the two tangent terms using the Pythagorean identity. Simplify the numerator and denominator.

$$\frac{\sec^3 x + 3\sec x - 4\tan^2 x - 4}{\tan^2 x - 4\sec x + 4} = \sec x$$
$$\frac{\sec^3 x + 3\sec x - 4\left(\sec^2 x - 1\right) - 4}{\left(\sec^2 x - 1\right) - 4\sec x + 4} =$$
$$\frac{\sec^3 x + 3\sec x - 4\sec^2 x + 4 - 4}{\sec^2 x - 1 - 4\sec x + 4} =$$
$$\frac{\sec^3 x - 4\sec^2 x + 3\sec x}{\sec^2 x - 4\sec x + 3} =$$

Now factor sec *x* out of each term in the numerator and reduce the fraction.

$$\frac{\sec x\left(\sec^2 x - 4\sec x + 3\right)^{1}}{{}_{1}\sec^2 x - 4\sec x + 3} =$$
$$\sec x = \sec x$$

7 Solve the identity by working on each side at the same time: $(\sin x + \cos x)\sin x = 1 - \cos x(\cos x - \sin x)$.

Distribute the terms on both sides. Then, on the left side, replace the first two terms using the appropriate Pythagorean identity.

$$(\sin x + \cos x)\sin x = 1 - \cos x(\cos x - \sin x)$$
$$\sin^2 x + \cos x \sin x = 1 - \cos^2 x + \cos x \sin x$$
$$= \left(1 - \cos^2 x\right) + \cos x \sin x$$
$$= \sin^2 x + \cos x \sin x$$

The two sides are now equal. Neither is the same as the beginning, but this is an equation — an identity.

8 Solve the identity by working on each side at the same time: $\frac{\sin 2x + \cos 2x}{\sin x \cos x} = 2 + \cot x - \tan x$.

On the left side, use the double angle identities to replace those terms. On the right, replace the cotangent and tangent using ratio identities, find a common denominator, and add the fractions.

$$\frac{\sin 2x + \cos 2x}{\sin x \cos x} = 2 + \cot x - \tan x$$
$$\frac{2\sin x \cos x + \cos^2 x - \sin^2 x}{\sin x \cos x} = 2 + \frac{\cos x}{\sin x} - \frac{\sin x}{\cos x}$$
$$= \frac{2\sin x \cos x}{\sin x \cos x} + \frac{\cos^2 x}{\sin x \cos x} - \frac{\sin^2 x}{\sin x \cos x}$$
$$= \frac{2\sin x \cos x + \cos^2 x - \sin^2 x}{\sin x \cos x}$$

9 Solve the identity by working on each side at the same time: $(\sin x + \cot x)^2 = \dfrac{\sin^4 x + \cos^2 x}{\sin^2 x} + 2\cos x$.

Square the binomial on the left. Split up the fraction on the right. Now, on the left, replace the cotangent using the ratio identity and simplify that term. On the right, reduce the first fraction and change the second to the cotangent on the left using the ratio identity.

$$(\sin x + \cot x)^2 = \frac{\sin^4 x + \cos^2 x}{\sin^2 x} + 2\cos x$$

$$\sin^2 x + 2\sin x \cot x + \cot^2 x = \frac{\sin^4 x}{\sin^2 x} + \frac{\cos^2 x}{\sin^2 x} + 2\cos x$$

$$\sin^2 x + 2\cancel{\sin x}\frac{\cos x}{\cancel{\sin x}} + \cot^2 x = \sin^2 x + \frac{\cos^2 x}{\sin^2 x} + 2\cos x$$

$$\sin^2 x + 2\cos x + \cot^2 x = \sin^2 x + \cot^2 x + 2\cos x$$

 Solve the identity by working on each side at the same time: $\dfrac{\sin(x-\pi)}{\sin(x+\pi)} = 1 - 2\sin^2 x - 2\cos^2 x$.

Apply the sum and difference formulas on the left. Factor out the –2, on the right, and apply the Pythagorean identity.

$$\frac{\sin(x-\pi)}{\sin(x+\pi)} = 3 - 2\sin^2 x - 2\cos^2 x$$

$$\frac{\sin x \cos\pi - \cos x \sin\pi}{\sin x \cos\pi + \cos x \sin\pi} = 3 - 2(\sin^2 x + \cos^2 x)$$

$$\frac{\sin x(-1) - \cos x(0)}{\sin x(-1) + \cos x(0)} = 3 - 2(1)$$

$$\frac{-\sin x}{-\sin x} = 1$$

11 Solve the identity by changing all the functions to sines and cosines: $\sec^2 x + \csc^2 x = \sec^2 x \csc^2 x$.

Work on only one side — I've chosen the left side. Change the two terms using reciprocal identities. Then find a common denominator and add them together. The numerator of the fraction is equal to 1, using the Pythagorean identity. Then the two terms in the denominator can be changed, using the reciprocal identity.

$$\sec^2 x + \csc^2 x = \sec^2 x \csc^2 x$$

$$\frac{1}{\cos^2 x} + \frac{1}{\sin^2 x} =$$

$$\frac{1}{\cos^2 x} \cdot \frac{\sin^2 x}{\sin^2 x} + \frac{1}{\sin^2 x} \cdot \frac{\cos^2 x}{\cos^2 x} =$$

$$\frac{\sin^2 x + \cos^2 x}{\sin^2 x \cos^2 x} =$$

$$\frac{1}{\sin^2 x \cos^2 x} =$$

$$\csc^2 x \sec^2 x = \sec^2 x \csc^2 x$$

12 Solve the identity by changing all the functions to sines and cosines: $\frac{\cos x + \cot x}{\tan x + \sec x} = \cos x \cot x$.

Change all the functions in the fraction using reciprocal and ratio identities. Add the fractions in the numerator together and in the denominator together, after finding common denominators. Multiply the numerator times the reciprocal of the denominator, and reduce. The ratio identity is used, at the end, to change the factor to cotangent.

$$\frac{\cos x + \frac{\cos x}{\sin x}}{\frac{\sin x}{\cos x} + \frac{1}{\cos x}} = \cos x \cot x$$

$$\frac{\frac{\cos x \sin x}{\sin x} + \frac{\cos x}{\sin x}}{\frac{\sin x}{\cos x} + \frac{1}{\cos x}} =$$

$$\frac{\frac{\cos x \sin x + \cos x}{\sin x}}{\frac{\sin x + 1}{\cos x}} =$$

$$\frac{\cos x \sin x + \cos x}{\sin x} \cdot \frac{\cos x}{\sin x + 1} =$$

$$\frac{\cos x \cancel{(\sin x + 1)}}{\sin x} \cdot \frac{\cos x}{\cancel{\sin x + 1}} =$$

$$\frac{\cos x}{\sin x} \cdot \cos x = \cot x \cos x$$

13 Solve the identity by changing all the functions to sines and cosines: $\sec^2 x + \cot^2 x = \csc^2 x + \tan^2 x$.

This identity is easier working on both sides — as well as changing everything to sine and cosine. Use the ratio and reciprocal identities, first. Then add the fractions together, on each side of the equation.

$$\sec^2 x + \cot^2 x = \csc^2 x + \tan^2 x$$

$$\frac{1}{\cos^2 x} + \frac{\cos^2 x}{\sin^2 x} = \frac{1}{\sin^2 x} + \frac{\sin^2 x}{\cos^2 x}$$

$$\frac{1}{\cos^2 x} \cdot \frac{\sin^2 x}{\sin^2 x} + \frac{\cos^2 x}{\sin^2 x} \cdot \frac{\cos^2 x}{\cos^2 x} = \frac{1}{\sin^2 x} \cdot \frac{\cos^2 x}{\cos^2 x} + \frac{\sin^2 x}{\cos^2 x} \cdot \frac{\sin^2 x}{\sin^2 x}$$

$$\frac{\sin^2 x}{\sin^2 x \cos^2 x} + \frac{\cos^4 x}{\sin^2 x \cos^2 x} = \frac{\cos^2 x}{\sin^2 x \cos^2 x} + \frac{\sin^4 x}{\sin^2 x \cos^2 x}$$

$$\frac{\sin^2 x + \cos^4 x}{\sin^2 x \cos^2 x} = \frac{\cos^2 x + \sin^4 x}{\sin^2 x \cos^2 x}$$

Write each fourth power as the product of two squares. Then apply the Pythagorean identity to each and distribute.

$$\frac{\sin^2 x + \cos^2 x \cos^2 x}{\sin^2 x \cos^2 x} = \frac{\cos^2 x + \sin^2 x \sin^2 x}{\sin^2 x \cos^2 x}$$

$$\frac{\sin^2 x + \cos^2 x (1 - \sin^2 x)}{\sin^2 x \cos^2 x} = \frac{\cos^2 x + \sin^2 x (1 - \cos^2 x)}{\sin^2 x \cos^2 x}$$

$$\frac{\sin^2 x + \cos^2 x - \sin^2 x \cos^2 x}{\sin^2 x \cos^2 x} = \frac{\cos^2 x + \sin^2 x - \sin^2 x \cos^2 x}{\sin^2 x \cos^2 x}$$

14 Solve the identity by changing all the functions to sines and cosines: $\frac{1+\cot x}{\cot x} = \tan x + \csc^2 x - \cot^2 x$.

This identity may look familiar. It's the same as Problem 2, but with a different technique. First, change every term on the right to its equivalent using ratio and reciprocal identities. Then find a common denominator and add all the fractions together.

$$\begin{aligned}
\frac{1+\cot x}{\cot x} &= \tan x + \csc^2 x - \cot^2 \\
&= \frac{\sin x}{\cos x} + \frac{1}{\sin^2 x} - \frac{\cos^2 x}{\sin^2 x} \\
&= \frac{\sin x}{\cos x}\cdot\frac{\sin^2 x}{\sin^2 x} + \frac{1}{\sin^2 x}\cdot\frac{\cos x}{\cos x} - \frac{\cos^2 x}{\sin^2 x}\cdot\frac{\cos x}{\cos x} \\
&= \frac{\sin^3 x}{\sin^2 x\cos x} + \frac{\cos x}{\sin^2 x\cos x} - \frac{\cos^3 x}{\sin^2 x\cos x} \\
&= \frac{\sin^3 x + \cos x - \cos^3 x}{\sin^2 x\cos x}
\end{aligned}$$

Now factor a cosine out of the last two terms in the numerator. Replace what's in the parentheses using the Pythagorean identity. Now divide each term in the fraction by the cube of the sine. Reduce, simplify, and apply the ratio identity.

$$\begin{aligned}
&= \frac{\sin^3 x + \cos x\left(1-\cos^2 x\right)}{\sin^2 x\cos x} \\
&= \frac{\sin^3 x + \cos x\left(\sin^2 x\right)}{\sin^2 x\cos x} \\
&= \frac{\frac{\sin^3 x}{\sin^3 x} + \frac{\cos x\left(\sin^2 x\right)}{\sin^3 x}}{\frac{\sin^2 x\cos x}{\sin^3 x}} \\
&= \frac{\frac{\cancel{\sin^3 x}}{\cancel{\sin^3 x}} + \frac{\cos x\left(\cancel{\sin^2 x}\right)}{\cancel{\sin^3} x}}{\frac{\cancel{\sin^2 x}\cos x}{\cancel{\sin^3} x}} \\
&= \frac{1+\frac{\cos x}{\sin x}}{\frac{\cos x}{\sin x}} \\
&= \frac{1+\cot x}{\cot x}
\end{aligned}$$

15 Solve the identity by multiplying by a conjugate: $\frac{\cot x}{\csc x - 1} = \frac{\csc x + 1}{\cot x}$.

Multiply the fraction on the left by the conjugate of the denominator. Then replace the square of the cosecant using a Pythagorean identity. Reduce the fraction.

$$\begin{aligned}
\frac{\cot x}{\csc x - 1} &= \frac{\csc x + 1}{\cot x} \\
\frac{\cot x}{\csc x - 1}\cdot\frac{\csc x + 1}{\csc x + 1} &= \\
\frac{\cot x\left(\csc x + 1\right)}{\csc^2 x - 1} &= \\
\frac{\cot x\left(\csc x + 1\right)}{1 + \cot^2 x - 1} &= \\
\frac{\cancel{\cot x}\left(\csc x + 1\right)}{\cot^{\cancel{2}} x} &= \\
\frac{\csc x + 1}{\cot x} &= \frac{\csc x + 1}{\cot x}
\end{aligned}$$

16 Solve the identity by multiplying by a conjugate: $\frac{\sin x}{1+\cos x} = \csc x - \cot x$.

Multiply the fraction on the left by the conjugate of the denominator. Substitute in the denominator using the Pythagorean identity. Then reduce the fraction. Separate the result into two separate fractions and apply the reciprocal and ratio identities.

$$\frac{\sin x}{1+\cos x} = \csc x - \cot x$$
$$\frac{\sin x}{1+\cos x} \cdot \frac{1-\cos x}{1-\cos x} =$$
$$\frac{\sin x(1-\cos x)}{1-\cos^2 x} =$$
$$\frac{\sin x(1-\cos x)}{\sin^2 x} =$$
$$\frac{{}^{1}\cancel{\sin x}(1-\cos x)}{\sin^{\cancel{2}1} x} =$$
$$\frac{1}{\sin x} - \frac{\cos x}{\sin x} =$$
$$\csc x - \cot x = \csc x - \cot x$$

17 Solve the identity by multiplying by a conjugate: $\frac{1-\sin x}{\cos x} = \frac{\cos x}{1+\sin x}$.

Multiply the fraction on the left by the conjugate of the numerator. Replace the numerator with its Pythagorean equivalent and reduce the fraction.

$$\frac{1-\sin x}{\cos x} = \frac{\cos x}{1+\sin x}$$
$$\frac{1-\sin x}{\cos x} \cdot \frac{1+\sin x}{1+\sin x} =$$
$$\frac{1-\sin^2 x}{\cos x(1+\sin x)} =$$
$$\frac{\cos^{\cancel{2}1} x}{{}_{1}\cancel{\cos x}(1+\sin x)} =$$
$$\frac{\cos x}{1+\sin x} = \frac{\cos x}{1+\sin x}$$

18 Solve the identity by multiplying by a conjugate: $\frac{1+\sin x+\cos x}{1+\sin x-\cos x} = \frac{1+\cos x}{\sin x}$.

This identity takes a careful choice of which conjugate to use. Take the fraction on the left, and break up the terms with the 1, first, and the next two terms as a grouping. Then multiply by the conjugate of the denominator.

$$\frac{1+(\sin x+\cos x)}{1+(\sin x-\cos x)} = \frac{1+\cos x}{\sin x}$$
$$\frac{1+(\sin x+\cos x)}{1+(\sin x-\cos x)} \cdot \frac{1-(\sin x-\cos x)}{1-(\sin x-\cos x)} =$$

When multiplying these fractions together, look at them as being binomials, and use FOIL.

$$\frac{1-(\sin x-\cos x)+(\sin x+\cos x)-(\sin x+\cos x)(\sin x-\cos x)}{1-(\sin x-\cos x)+(\sin x-\cos x)-(\sin x-\cos x)(\sin x-\cos x)} =$$
$$\frac{1-\sin x+\cos x+\sin x+\cos x-(\sin^2 x-\cos^2 x)}{1-\sin x+\cos x+\sin x-\cos x-(\sin^2 x-2\sin x\cos x+\cos^2 x)} =$$
$$\frac{1-\sin x+\cos x+\sin x+\cos x-\sin^2 x+\cos^2 x}{1-\sin x+\cos x+\sin x-\cos x-\sin^2 x+2\sin x\cos x-\cos^2 x} =$$

Simplify the fraction, combining like terms. Replace the Pythagorean expression in the denominator with a 1 and the sine squared term in the numerator with its Pythagorean equivalent.

$$\frac{1+2\cos x-(1-\cos^2 x)+\cos^2 x}{1+2\sin x\cos x-(\sin^2 x+\cos^2 x)}=$$

$$\frac{1+2\cos x-1+\cos^2 x+\cos^2 x}{1+2\sin x\cos x-(1)}=$$

$$\frac{2\cos x+2\cos^2 x}{2\sin x\cos x}=$$

Now factor the numerator and reduce the fraction.

$$\frac{2\cos x(1+\cos x)}{2\sin x\cos x}=$$

$$\frac{{}^{1}\cancel{2}\cancel{\cos x}(1+\cos x)}{{}_{1}\cancel{2}\sin x\cancel{\cos x}}=$$

$$\frac{1+\cos x}{\sin x}=\frac{1+\cos x}{\sin x}$$

19 Solve the identity $\sqrt{\frac{1-\sin x}{1+\sin x}}=\frac{\cos x}{1+\sin x}$.

Square both sides. Replace the numerator on the right using a Pythagorean identity. Then factor the numerator and reduce the fraction.

$$\sqrt{\frac{1-\sin x}{1+\sin x}}=\frac{\cos x}{1+\sin x}$$

$$\left(\sqrt{\frac{1-\sin x}{1+\sin x}}\right)^2=\left(\frac{\cos x}{1+\sin x}\right)^2$$

$$\frac{1-\sin x}{1+\sin x}=\frac{\cos^2 x}{(1+\sin x)^2}$$

$$=\frac{1-\sin^2 x}{(1+\sin x)^2}$$

$$=\frac{(1-\sin x)\cancel{(1+\sin x)}^{1}}{(1+\sin x)^{\cancel{2}1}}$$

$$=\frac{1-\sin x}{1+\sin x}$$

20 Solve the identity $\sqrt{\frac{1+\cos x}{1-\cos x}}=\csc x+\cot x$.

Square both sides. Multiply out on the right. Then change each term on the right, using reciprocal and ratio identities. They all have the same denominator, so add the fractions.

$$\sqrt{\frac{1+\cos x}{1-\cos x}}=\csc x+\cot x$$

$$\left(\sqrt{\frac{1+\cos x}{1-\cos x}}\right)^2=(\csc x+\cot x)^2$$

$$\frac{1+\cos x}{1-\cos x}=\csc^2 x+2\csc x\cot x+\cot^2 x$$

$$=\frac{1}{\sin^2 x}+\frac{2}{\sin x}\frac{\cos x}{\sin x}+\frac{\cos^2 x}{\sin^2 x}$$

$$=\frac{1+2\cos x+\cos^2 x}{\sin^2 x}$$

Replace the denominator using a Pythagorean identity. Then factor it. Now factor the numerator — it's a perfect square binomial. Reduce the fraction.

$$= \frac{1 + 2\cos x + \cos^2 x}{1 - \cos^2 x}$$

$$= \frac{(1 + \cos x)^{\not 2\,1}}{_1\cancel{(1 + \cos x)}(1 - \cos x)}$$

$$\frac{1 + \cos x}{1 - \cos x} = \frac{1 + \cos x}{1 - \cos x}$$

21 Solve the identity: $\frac{1}{\sin x + 1} - \frac{1}{\sin x - 1} = 2\sec^2 x$.

Add the two fractions on the left together, after finding their common denominator and changing both fractions. Simplify the numerator. Replace the square of sine in the denominator using a Pythagorean identity. Then finish it up with the reciprocal identity for secant.

$$\frac{1}{\sin x + 1} \cdot \frac{\sin x - 1}{\sin x - 1} - \frac{1}{\sin x - 1} \cdot \frac{\sin x + 1}{\sin x + 1} = 2\sec^2 x$$

$$\frac{\sin x - 1}{\sin^2 x - 1} - \frac{\sin x + 1}{\sin^2 x - 1} =$$

$$\frac{\sin x - 1 - (\sin x + 1)}{\sin^2 x - 1} =$$

$$\frac{-2}{\sin^2 x - 1} =$$

$$\frac{-2}{(1 - \cos^2 x) - 1} =$$

$$\frac{-2}{-\cos^2 x} =$$

$$2\sec^2 x = 2\sec^2 x$$

22 Solve the identity: $\frac{\sec x}{\sin x} + \frac{\csc x}{\cos x} = \frac{4}{\sin 2x}$.

Find a common denominator, and add the two fractions on the left together. Notice that you're multiplying two reciprocals together in each numerator, so the product is 1. Multiply numerator and denominator of the fraction by 2. Replace the denominator with its double angle equivalent.

$$\frac{\sec x}{\sin x} + \frac{\csc x}{\cos x} = \frac{4}{\sin 2x}$$

$$\frac{\sec x}{\sin x} \cdot \frac{\cos x}{\cos x} + \frac{\csc x}{\cos x} \cdot \frac{\sin x}{\sin x} =$$

$$\frac{1}{\sin x \cos x} + \frac{1}{\sin x \cos x} =$$

$$\frac{2}{\sin x \cos x} =$$

$$\frac{4}{2\sin x \cos x} =$$

$$\frac{4}{\sin 2x} = \frac{4}{\sin 2x}$$

23 Solve the identity: $\frac{1}{\sec x - 1} - \frac{1}{\sec x + 1} = 2\cot^2 x$.

Find a common denominator, and add the two fractions together. Replace the square of secant in the denominator with its equivalent, using a Pythagorean identity. Simplify, and apply a reciprocal identity.

$$\frac{1}{\sec x - 1} - \frac{1}{\sec x + 1} = 2\cot^2 x$$

$$\frac{1}{\sec x - 1} \cdot \frac{\sec x + 1}{\sec x + 1} - \frac{1}{\sec x + 1} \cdot \frac{\sec x - 1}{\sec x - 1} =$$

$$\frac{\sec x + 1}{\sec^2 x - 1} - \frac{\sec x - 1}{\sec^2 x - 1} =$$

$$\frac{\sec x + 1 - (\sec x - 1)}{\sec^2 x - 1} =$$

$$\frac{2}{\sec^2 x - 1} =$$

$$\frac{2}{(\tan^2 x + 1) - 1} =$$

$$\frac{2}{\tan^2 x} =$$

$$2\cot^2 x = 2\cot^2 x$$

24 Solve the identity: $\frac{\sin x}{1 + \cos x} + \frac{1 + \cos x}{\sin x} = 2\csc x$.

Find a common denominator, and add the two fractions together. Square the binomial in the numerator. Then combine the two terms from the Pythagorean identity to get 1. Factor the numerator, and then reduce the fraction. Use the reciprocal identity to finish the job.

$$\frac{\sin x}{1 + \cos x} + \frac{1 + \cos x}{\sin x} = 2\csc x$$

$$\frac{\sin x}{1 + \cos x} \cdot \frac{\sin x}{\sin x} + \frac{1 + \cos x}{\sin x} \cdot \frac{1 + \cos x}{1 + \cos x} =$$

$$\frac{\sin^2 x}{\sin x(1 + \cos x)} + \frac{(1 + \cos x)^2}{\sin x(1 + \cos x)} =$$

$$\frac{\sin^2 x + (1 + \cos x)^2}{\sin x(1 + \cos x)} =$$

$$\frac{\sin^2 x + 1 + 2\cos x + \cos^2 x}{\sin x(1 + \cos x)} =$$

$$\frac{(\sin^2 x + \cos^2 x) + 1 + 2\cos x}{\sin x(1 + \cos x)} =$$

$$\frac{2 + 2\cos x}{\sin x(1 + \cos x)} =$$

$$\frac{2\cancel{(1 + \cos x)}^1}{\sin x\cancel{(1 + \cos x)}_1} =$$

$$\frac{2}{\sin x} = 2\csc x$$

25 Write the cosine function in terms of sine. $\cos x = \sqrt{1-\sin^2 x}$.

Start with the Pythagorean identity and solve for cosine.

$$\begin{aligned}\sin^2 x + \cos^2 x &= 1\\ \cos^2 x &= 1 - \sin^2 x\\ \sqrt{\cos^2 x} &= \sqrt{1-\sin^2 x}\\ \cos x &= \sqrt{1-\sin^2 x}\end{aligned}$$

26 Write the cosine function in terms of tangent. $\cos x = \dfrac{1}{\sqrt{\tan^2 x + 1}}$.

Start with the Pythagorean identity involving tangent and secant. The secant is the reciprocal of cosine, so that will be used in the process. After replacing the secant with cosine, solve for the cosine by flipping the proportion.

$$\begin{aligned}\tan^2 x + 1 &= \sec^2 x\\ \tan^2 x + 1 &= \frac{1}{\cos^2 x}\\ \frac{\tan^2 x + 1}{1} &= \frac{1}{\cos^2 x}\\ \frac{1}{\tan^2 x + 1} &= \frac{\cos^2 x}{1}\\ \sqrt{\cos^2 x} &= \sqrt{\frac{1}{\tan^2 x + 1}}\\ \cos x &= \frac{1}{\sqrt{\tan^2 x + 1}}\end{aligned}$$

27 Write the tangent function in terms of sine. $\tan x = \dfrac{\sin x}{\sqrt{1-\sin^2 x}}$.

Start with the ratio identity. Then replace the cosine with its equivalence from Problem 25.

$$\begin{aligned}\tan x &= \frac{\sin x}{\cos x}\\ &= \frac{\sin x}{\sqrt{1-\sin^2 x}}\end{aligned}$$

28 Write the secant function in terms of cosecant. $\sec x = \dfrac{\csc x}{\sqrt{\csc^2 x - 1}}$.

Start with the Pythagorean identity involving sine and cosine, the reciprocals of these two functions. Replace the sine and cosine with reciprocal equivalences. Then solve for secant.

$$\begin{aligned}\sin^2 x + \cos^2 x &= 1\\ \frac{1}{\csc^2 x} + \frac{1}{\sec^2 x} &= 1\\ \frac{1}{\sec^2 x} &= 1 - \frac{1}{\csc^2 x}\\ \frac{1}{\sec^2 x} &= \frac{\csc^2 x}{\csc^2 x} - \frac{1}{\csc^2 x}\\ \frac{1}{\sec^2 x} &= \frac{\csc^2 x - 1}{\csc^2 x}\\ \frac{\sec^2 x}{1} &= \frac{\csc^2 x}{\csc^2 x - 1}\\ \sqrt{\sec^2 x} &= \sqrt{\frac{\csc^2 x}{\csc^2 x - 1}}\\ \sec x &= \frac{\csc x}{\sqrt{\csc^2 x - 1}}\end{aligned}$$

Chapter 12

Introducing Inverse Trig Functions

In This Chapter

- Figuring out where inverse functions go
- Using inverse trig functions in equations
- Finding the single solution or the many solutions

Inverse trig functions work much like the inverses of the algebraic functions. The inverse reverses the operations and figures out what the input was. The main difference between the algebraic and trig inverses, though, is that there are restrictions on the results of applying the inverse trig functions. Each function has its own designation in terms of which quadrants apply.

There are inverse trig functions and inverse trig relations. The functions act as all functions do — one answer for each input. The relations give an infinite number of answers — everything that works. For example, the inverse sine *function* is designated with $\text{Arcsin}x$ or $\text{Sin}^{-1}x$, and the inverse sine *relation* is designated with $\arcsin x$ or $\sin^{-1}x$. The rest of the functions have similar notation. To tell them apart, the inverse trig functions are capitalized, and the relations aren't.

The –1 superscript written between the function and the angle means that this is an inverse function; it doesn't mean that the function is being raised to that power.

Determining the Correct Quadrants

When applying the inverse sine function, the *range* (output) or results are always angles in Quadrant I and Quadrant IV. The same is true of the cosecant and tangent. When applying the inverse cosine, secant, or cotangent, the answer is always an angle in Quadrant I or II. This way, there's only one possible answer for each inverse function input.

Q. If you input a *positive* number into the inverse cosine function, in which quadrant will the answer lie?

A. Quadrant I. The cosine is positive in Quadrant I and in Quadrant IV, but the inverse cosine function doesn't have results in Quadrant IV.

Q. If you input a *negative* number into the inverse tangent function, in which quadrant will the answer lie?

A. Quadrant IV. The tangent is negative in Quadrant II and Quadrant IV, but the inverse tangent function doesn't have results in Quadrant II.

1. Which inverse trig functions have range values or results in Quadrant I?

2. Which inverse trig functions have range values or results in Quadrant II?

Solve It

3. Which inverse trig functions have range values or results in Quadrant III?

4. Which inverse trig functions have range values or results in Quadrant IV?

Solve It

Evaluating Expressions Using Inverse Trig Functions

The notation for the inverse trig functions involves either a –1 superscript or the word *Arc* in front of the name. In either case, the inverse function finds the angle measure that is responsible for the particular value being input. The angle measure is taken from the correct quadrant, as described in the previous section.

Q. Evaluate $\text{Cos}^{-1}\left(\frac{1}{2}\right)$.

A. 60 degrees. The input value is positive, so the answer has to be from Quadrant I, where the cosine is positive. The cosine of 60 degrees is equal to ½.

Q. Evaluate $\text{Arcsec}\left(-\frac{2\sqrt{3}}{3}\right)$.

A. 150 degrees. The input value is negative, so the answer has to be from Quadrant II, where the secant is negative. You can get this value from the table or you can use reference angles and quadrants. The secant of 30 degrees is $\frac{2\sqrt{3}}{3}$. The 30-degree angle is the reference angle for 150 degrees in Quadrant II.

5. Evaluate $\text{Sin}^{-1}\left(-\frac{1}{2}\right)$.

Solve It

6. Evaluate $\text{Arccos}\left(\frac{\sqrt{3}}{2}\right)$.

Solve It

7. Evaluate $\text{Arctan}\left(-\sqrt{3}\right)$.

Solve It

8. Evaluate $\text{Cot}^{-1}(-1)$.

Solve It

9. Evaluate $\text{Sec}^{-1}(2)$.

Solve It

10. Evaluate $\text{Arccsc}\left(-\sqrt{2}\right)$.

Solve It

Solving Equations Using Inverse Trig Functions

The whole point of having these inverse trig functions is to put them to use when solving equations involving trig functions. You start out with an equation that contains one or more trig functions, apply any identities or algebraic operations necessary, and then write the solution in terms of an inverse trig function — and evaluate that function.

The examples here and the practice problems for this section apply the inverse function property or definition: There's one answer for each input. Sections later in this chapter deal with some other cases — where there could be more than one answer to an equation.

EXAMPLE

Q. Solve the equation $\sin x + \frac{1}{2} = 0$.

A. 330 degrees. First solve for sin x by subtracting ½ from each side. Then rewrite the equation, solving for x.

$$\sin x + \frac{1}{2} = 0$$
$$\sin x = -\frac{1}{2}$$
$$x = \text{Sin}^{-1}\left(-\frac{1}{2}\right)$$

This last line is read, "x is the angle whose sine is negative one-half." Using the inverse sine function, the answer is 330 degrees. You get this value from the table in the appendix, or you can use the reference angle of 30 degrees in Quadrant IV to reason this out.

Q. Solve the equation $\tan x \sin x - \tan x = 0$.

A. 0 degrees and 90 degrees. First, factor out the tangent. Then set each of the factors equal to 0 and solve for x in each.

$$\tan x(\sin x - 1) = 0$$
$$\tan x = 0 \qquad \sin x - 1 = 0$$
$$\sin x = 1$$
$$x = \text{Tan}^{-1}(0) \qquad x = \text{Sin}^{-1}(1)$$

In the equation involving the inverse tangent, the answer is $x = 0$ degrees. This angle lies between Quadrant I and Quadrant IV, where the inverse tangent is defined. In the equation involving the inverse sine, the answer is $x = 90$ degrees. This angle lies between Quadrant I and Quadrant II, where the inverse sine is defined.

11. Solve the equation $\sin x = \frac{\sqrt{2}}{2}$.

Solve It

12. Solve the equation $\sin x \cos x - \cos x = 0$.

Solve It

13. Solve the equation $\tan^2 x - 1 = 0$.

14. Solve the equation $\sec^2 x - 2\sec x = 0$.

15. Solve the equation $2\cot x \sin^2 x - \sin x \cot x = 0$.

16. Solve the equation $4\csc^2 x - 1 = 0$.

Creating Multiple Answers for Multiple and Half-Angles

The trig functions are *periodic,* which means that they repeat the same values over and over again in a predictable pattern. Knowing when a particular value will occur if you're plotting temperatures or seasonal sales is helpful. Sometimes, more than one answer is needed.

One trig relation can have an infinite number of answers. For instance, in $x = \sin^{-1}\left(\frac{1}{2}\right)$, $x = 30, 150, 390, 510, 750, 870, \ldots$ You can get pretty tired listing all of them. There is a better way that uses a rule and the letter k to represent all the *integers* (positive and negative whole numbers and 0).

Q. List all the solutions of $x = \sin^{-1}\left(\frac{1}{2}\right)$ in degrees.

A. $x = 30 + 360k$ or $x = 150 + 360k$. These two rules will generate all the possible answers. If $k = 0$, you just have the two angles 30 and 150. If $k = 1$, you have $30 + 360 = 390$ and $150 + 360 = 410$. Any integer will give you another answer.

Q. List all the solutions of $x = \tan^{-1}(1)$ in radians.

A. $x = \frac{\pi}{4} + k\pi$. Only one rule is needed here, because the positive tangent values are in quadrants that are diagonal from one another (as are the negative).

17. List all the solutions of $x = \cos^{-1}\left(-\frac{\sqrt{3}}{2}\right)$ in degrees.

Solve It

18. List all the solutions of $x = \sin^{-1}(0)$ in degrees.

19. List all the solutions of $x = \tan^{-1}\left(-\sqrt{3}\right)$ in degrees.

Solve It

20. List all the solutions of $x = \sec^{-1}(2)$ in radians.

Solve It

21. List all the solutions of $x = \cot^{-1}(-1)$ in radians.

Solve It

22. List all the solutions of $x = \cos^{-1}(-1)$ in radians.

Solve It

Answers to Problems on Inverse Trig Functions

The following are solutions to the practice problems presented earlier in this chapter.

1 Which inverse trig functions have range values or results in Quadrant I? **All the inverse trig functions have their values in Quadrant I.**

They're all defined there.

2 Which inverse trig functions have range values or results in Quadrant II? **The inverse cosine, secant, and cotangent have their values in Quadrant II.**

These functions are negative in Quadrant II.

3 Which inverse trig functions have range values or results in Quadrant III? **None of the inverse trig functions are defined for Quadrant III.**

4 Which inverse trig functions have range values or results in Quadrant IV? **The inverse sine, cosecant, and tangent have values in Quadrant IV.**

They're all negative in this quadrant.

5 Evaluate $\text{Sin}^{-1}\left(-\frac{1}{2}\right)$. **330 degrees (or $\frac{11\pi}{6}$ in radians).**

6 Evaluate $\text{Arccos}\left(\frac{\sqrt{3}}{2}\right)$. **30 degrees (or $\frac{\pi}{6}$ in radians).**

7 Evaluate $\text{Arctan}\left(-\sqrt{3}\right)$. **300 degrees (or $\frac{5\pi}{3}$ in radians).**

8 Evaluate $\text{Cot}^{-1}(-1)$. **135 degrees (or $\frac{3\pi}{4}$ in radians).**

9 Evaluate $\text{Sec}^{-1}(2)$. **60 degrees (or $\frac{\pi}{3}$ in radians).**

10 Evaluate $\text{Arccsc}\left(-\sqrt{2}\right)$. **315 degrees (or $\frac{7\pi}{4}$ in radians).**

11 Solve the equation $\sin x = \frac{\sqrt{2}}{2}$. $\boldsymbol{x = \text{Sin}^{-1}\left(\frac{\sqrt{2}}{2}\right) = 45}$ **degrees (or $\frac{\pi}{4}$ in radians).**

12 Solve the equation $\sin x \cos x - \cos x = 0$. **90 degrees (or $\frac{\pi}{2}$ in radians).**

Factor out cos x. Then set each factor equal to 0 and solve for x.

$$\cos x(\sin x - 1) = 0$$

$$\begin{array}{ll} \cos x = 0 & \sin x - 1 = 0 \\ x = \text{Cos}^{-1}(0) = 90 & \sin x = 1 \\ & x = \text{Sin}^{-1}(1) = 90 \end{array}$$

The factors each yield the same answer. This is the only solution.

13 Solve the equation $\tan^2 x - 1 = 0$. **45 degrees or 315 degrees (or $\frac{\pi}{4}$ or $\frac{7\pi}{4}$ in radians).**

Factor first. Then set each factor equal to 0 and solve for x.

$$(\tan x - 1)(\tan x + 1) = 0$$

$$\begin{array}{ll} \tan x - 1 = 0 & \tan x + 1 = 0 \\ \tan x = 1 & \tan x = -1 \\ x = \text{Tan}^{-1}(1) = 45 & x = \text{Tan}^{-1}(-1) = 315 \end{array}$$

14 Solve the equation $\sec^2 x - 2\sec x = 0$. **60 degrees (or $\frac{\pi}{3}$ in radians).**

Factor out sec x. Then set each factor equal to 0 and solve for x.

$$\sec x(\sec x - 2) = 0$$
$$\sec x = 0 \quad \sec x - 2 = 0$$
$$* \quad \sec x = 2$$
$$x = \mathrm{Sec}^{-1}(2) = 60$$

The first part has no solution — that's why the *. The secant function has values 1 and bigger or –1 and smaller — nothing in between.

15 Solve the equation $2\cot x \sin^2 x - \sin x \cot x = 0$. **0, 30, 90 degrees (or 0, $\frac{\pi}{6}$, $\frac{\pi}{2}$ in radians).**

Factor out sin *x* cot *x*. Then set each of the three factors equal to 0 and solve for *x*.

$$\sin x \cot x(2\sin x - 1) = 0$$
$$\sin x = 0 \quad \cot x = 0 \quad 2\sin x - 1 = 0$$
$$x = \mathrm{Sin}^{-1}(0) = 0 \quad x = \mathrm{Cot}^{-1}(0) = 90 \quad \sin x = \frac{1}{2}$$
$$x = \mathrm{Sin}^{-1}\left(\frac{1}{2}\right) = 30$$

16 Solve the equation $4\csc^2 x - 1 = 0$. **No solution.**

First factor; then set the factors equal to 0 and solve for *x*.

$$(2\csc x - 1)(2\csc x + 1) = 0$$
$$2\csc x - 1 = 0 \quad 2\csc x + 1 = 0$$
$$\csc x = \frac{1}{2} \quad \csc x = -\frac{1}{2}$$
$$* \quad *$$

There is no solution — that's why the *. The cosecant function has values 1 and bigger or –1 and smaller — nothing in between.

17 List all the solutions of $x = \cos^{-1}\left(-\frac{\sqrt{3}}{2}\right)$ in degrees. **x = 150 + 360k or x = 210 + 360k.**

The two solutions within one complete revolution are x = 150 and x = 210. Add multiples of 360 to each.

18 List all the solutions of $x = \sin^{-1}(0)$ in degrees. **x = 180k.**

The two solutions within one complete revolution are x = 0 and x = 180. Adding 180 to 0 gives you 180, and adding 180 repeatedly gives you all the values.

19 List all the solutions of $x = \tan^{-1}(-\sqrt{3})$ in degrees. **x = 120 + 180k.**

The solutions within one complete revolution are x = 120 and x = 300. Adding 180 to 120 gives you 300. All the solutions are obtained by adding multiples of 180 to 120.

20 List all the solutions of $x = \sec^{-1}(2)$ in radians. **$x = \frac{\pi}{3} + 2k\pi$ or $x = \frac{5\pi}{3} + 2k\pi$.**

The solutions within one complete revolution are $x = \frac{\pi}{3}$ and $x = \frac{5\pi}{3}$. Adding multiples of 2π to each of those gives you all the solutions.

21 List all the solutions of $x = \cot^{-1}(-1)$ in radians. **$x = \frac{3\pi}{4} + k\pi$.**

The solutions within one complete revolution are $\frac{3\pi}{4}$ and $\frac{7\pi}{4}$. They're just π apart, so all the solutions can be written by just adding multiples of π to $\frac{3\pi}{4}$.

22 List all the solutions of $x = \cos^{-1}(-1)$ in radians. **x = (2k + 1)π.**

The only solution within one complete revolution is **x = (2k + 1)π**. All the solutions are obtained with odd multiples of π.

Chapter 13

Solving Trig Equations

In This Chapter

- Solving for the few or the many angles
- Solving equations using factoring techniques
- Changing to fractions to solve equations

Solving trigonometric equations involves many of the same techniques used in solving algebraic equations, such as factoring, squaring both sides, and applying the quadratic formula. One of the things that makes solving trig equations special is the ability to insert completely different expressions by using the identities. Another thing that makes them different is that you can have multiple answers — mainly due to the periodic nature of the trig functions and the way their values repeat.

Solving for Solutions within One Rotation

One rotation in trigonometry means a sweep of 360 degrees or 2π radians. It's one time around. Many times, this is all you need — just the answers to a trig equation within one complete rotation. These answers usually come in pairs, because the same function values occur in two of the four quadrants. The sines of both 30-degree and 150-degree angles are ½.

Q. Solve for x between and including 0 and 360 degrees (written $0 \leq x \leq 360$) in $3\tan^2 x - 2\sqrt{3}\tan x - 3 = 0$.

A. $x = 60, 150, 240,$ and 330 degrees (or $\frac{\pi}{3}, \frac{5\pi}{6}, \frac{4\pi}{3}, \frac{11\pi}{6}$ radians). First, factor the quadratic. This particular factorization isn't all that obvious except for the fact that $\sqrt{3}$ does occur in many of the exact values for the more familiar angles.

$$\left(\tan x - \sqrt{3}\right)\left(3\tan x + \sqrt{3}\right) = 0$$

Now set each of the factors equal to 0 and determine, using inverse trig functions, which angles between 0 and 360 satisfy the equations.

$$\tan x - \sqrt{3} = 0 \qquad\qquad 3\tan x + \sqrt{3} = 0$$

$$\tan x = \sqrt{3} \qquad\qquad 3\tan x = -\sqrt{3}$$

$$\tan x = -\frac{\sqrt{3}}{3}$$

$$x = \tan^{-1}\left(\sqrt{3}\right) = 60,\ 240 \qquad\qquad x = \tan^{-1}\left(-\frac{\sqrt{3}}{3}\right) = 150,\ 330$$

This example is done in degrees, but it could have just as easily been done in terms of radians. It's your choice.

1. Solve for x in $2\sin^2 x - \sin x - 1 = 0$ when $0 \le x \le 360$.

2. Solve for x in $\cos^2 x - \frac{\sqrt{3}}{2}\cos x = 0$ when $0 \le x \le 360$.

Solve It

3. Solve for x in $\tan^2 x - 1 = 0$ when $0 \le x \le 360$.

Solve It

4. Solve for x in $\sec^2 - 4 = 0$ when $0 \le x \le 360$.

Solve It

5. Solve for x in $1 + \cot^2 x + 2\csc x = 0$ when $0 \le x \le 360$.

Solve It

6. Solve for x in $\sin 2x - \cos x = 0$ when $0 \le x \le 360$.

Solve It

Solving Equations with Multiple Answers

Trig equations with multiple angles are going to have multiple answers. The coefficient or multiplier on the variable is going to allow for more answers within one full rotation.

When solving equations for multiple angle values, there's a technique to follow:

1. **Solve for the angle measure, writing an equation involving an inverse trig *relation*.**

   This isn't a function; there's more than one answer.
2. **List all angles that satisfy this equation within one rotation.**
3. **Add 360 degrees (or 2π) to each angle measure, repeating for as many multiples as are indicated in the problem.**

   If the angle is $2x$, you want two rotations, so add the 360 once to each angle. If it's $3x$, add it twice, and so on.
4. **Divide each angle measure by the coefficient (multiplier) of the x.**

EXAMPLE

Q. Solve $\sin 3x = \frac{\sqrt{3}}{2}$.

A. 20, 40, 140, 160, 260, 280 degrees. First, solve for $3x$. Then determine the two angles within one rotation that have this function value. Notice that the inverse relation is indicated with a lowercase *s* to show it isn't the function.

$$3x = \sin^{-1}\left(\frac{\sqrt{3}}{2}\right)$$

$$3x = 60,\ 120$$

The sine is positive in Quadrants I and II, and the sine is equal to $\frac{\sqrt{3}}{2}$ for both of these angles. Now add 360 degrees to each angle, and then add 360 degrees to those two results. That gives you three full rotations of angles measuring $3x$ that have that sine. $3x = 60, 120, 420, 480, 780, 840$. Divide $3x$ and each of these angles by 3 to solve for x.

$$\frac{3x}{3} = \frac{60}{3}, \frac{120}{3}, \frac{420}{3}, \frac{480}{3}, \frac{780}{3}, \frac{840}{3}$$

$$x = 20,\ 40,\ 140,\ 160,\ 260,\ 280$$

Q. Solve for x in $2\sin^2 2x - \sin 2x = 0$ when $0 \le x \le 360$.

A. $x = 0, 15, 75, 90, 180, 195, 205, 270, 360$ degrees. The first thing to do is factor the equation. Then set the two factors equal to 0 and first determine what angles work for the angle $2x$.

$$\sin 2x\,(2\sin 2x - 1) = 0$$

$$\sin 2x = 0 \quad \text{or} \quad 2\sin 2x - 1 = 0$$

$$2\sin 2x = 1$$

$$\sin 2x = \frac{1}{2}$$

$$2x = 0,\ 180,\ 360,\ 540,\ 720 \qquad 2x = 30, 150, 390, 410$$

Note that I included 720 degrees, because it's at the end of the second full revolution. Now solve for x by dividing every angle by 2.

$$\frac{2x}{x} = \frac{0}{2}, \frac{180}{2}, \frac{360}{2}, \frac{540}{2}, \frac{720}{2} \qquad \frac{2x}{2} = \frac{30}{2}, \frac{150}{2}, \frac{390}{2}, \frac{410}{2}$$

$$x = 0,\ 90,\ 180,\ 270,\ 360 \qquad x = 15, 75, 195, 205$$

7. Solve for x in $\sin 2x = -\frac{1}{2}$.

Solve It

8. Solve for x in $\cos 3x = \frac{\sqrt{2}}{2}$.

Solve It

9. Solve for x in $\sec^2 2x - \sec 2x = 0$.

Solve It

10. Solve for x in $2\sin^2 3x - 3\sin 3x + 1 = 0$.

Solve It

Special Factoring for a Solution

The standard techniques of factoring out a *greatest common factor* (the largest value that will divide all the terms evenly) or using *unFOIL* (factoring into the product of two binomials) are the ones most commonly used when solving trigonometric equations. The types of factoring that also may be needed are factoring binomials that have two perfect cubes, factoring by grouping, and factoring quadratic-like equations. One that can be dispatched rather quickly is the one involving perfect cubes. Doing the factorization doesn't really gain you much, so you might as well just take the cube root of each side. I've included that little technique and explanation in this first example that involves factoring by grouping.

Q. Solve the equation $2\sin^3 x\cos x - \sin^3 x - 2\cos x + 1 = 0$.

A. $x = 60$, 90, and 300 degrees. Factor the equation using grouping — the first two terms have a common factor and the second two terms have a different common factor. Then factor the common binomial out of those two results.

$$\sin^3 x(2\cos x - 1) - 1(2\cos x - 1) = 0$$
$$(2\cos x - 1)(\sin^3 x - 1) = 0$$

The second factor here can be factored into $(\sin x - 1)(\sin^2 x + \sin x + 1)$, but the trinomial doesn't have any real solution, so doing the factoring doesn't provide any answer. Just move the 1 to the right and take the cube root of each side. The other factor can be solved for x as usual.

$$(2\cos x - 1)(\sin^3 x - 1) = 0$$

$$2\cos x - 1 = 0 \quad \text{or} \quad \sin^3 x - 1 = 0$$
$$2\cos x = 1 \qquad \sin^3 x = 1$$
$$\cos x = \frac{1}{2} \qquad \sqrt[3]{\sin^3 x} = \sqrt[3]{1}$$
$$\sin x = 1$$

$$x = \cos^{-1}\left(\frac{1}{2}\right) = 60,\ 300 \qquad x = \sin^{-1}(1) = 90$$

Q. Solve the equation $\sec^6 x - 9\sec^3 x + 8 = 0$.

A. $x = 0$, 60, 300, and 360 degrees. This is a quadratic-like equation. It's a trinomial where the highest power is twice the other power, and the third term is a constant. You try to factor it into the product of two binomials and then solve equations that are formed when those factors are set equal to 0. Here's the factorization, first: $(\sec^3 x - 1)(\sec^3 x - 8) = 0$. Set each factor equal to 0, and solve for the values of x that work.

$$\sec^3 x - 1 = 0 \qquad \sec^3 x - 8 = 0$$
$$\sqrt[3]{\sec^3 x} = \sqrt[3]{1} \qquad \sqrt[3]{\sec^3 x} = \sqrt[3]{8}$$
$$\sec x = 1 \qquad \sec x = 2$$
$$x = \sec^{-1}(1) = 0,\ 360 \qquad x = \sec^{-1}(2) = 60,\ 300$$

11. Solve for x in $\csc^3 x = 8$.

Solve It

12. Solve for x in $\tan^3 x + 1 = 0$.

Solve It

13. Solve for x in
$2\sin x \tan x + \tan x - 2\sqrt{3}\sin x - \sqrt{3} = 0$.

Solve It

14. Solve for x in
$\tan x \sec x - 2\tan x + \sec x - 2 = 0$.

Using Fractions and Common Denominators to Solve Equations

Fractions aren't everyone's cup of tea. People avoid them just because they're so fussy to deal with. What's the fuss about? Oh, it's those common denominators. In trig, though, the tables are turned and you seek out fractions as aids to solving identities and equations. When you have a trig problem with fractions that have to be added, you find a common denominator and then apply identities to finish up. And, even when you don't already have fractions, you introduce them by applying the identities, and then find the common denominator. It just works so nicely.

Q. Solve for x in $\sec x - 2\tan x = 0$ when $0 \le x \le 360$.

A. $x = 30$ or 150 degrees. This is a case where introducing fractions is to your benefit. Using a reciprocal and ratio identity, the two fractions have a common denominator cos x and can be subtracted to form one fraction.

$$\frac{1}{\cos x} - \frac{2\sin x}{\cos x} = 0$$
$$\frac{1 - 2\sin x}{\cos x} = 0$$

Now, setting the numerator equal to 0,

$$1 - 2\sin x = 0$$
$$1 = 2\sin x$$
$$\frac{1}{2} = \sin x$$
$$x = \sin^{-1}\left(\frac{1}{2}\right) = 30,\ 150$$

You don't set the denominator equal to 0 — that would create a number that doesn't exist. These two angles are the only solutions.

Q. Solve for x in $\csc x - \tan x + \csc x \sec x = 0$ when $0 \le x \le 360$.

A. There's no solution. First change the functions using reciprocal and ratio identities. Then find a common denominator, rewrite the fractions, and combine them.

$$\frac{1}{\sin x} - \frac{\sin x}{\cos x} + \frac{1}{\sin x \cos x} = 0$$
$$\frac{1}{\sin x} \cdot \frac{\cos x}{\cos x} - \frac{\sin x}{\cos x} \cdot \frac{\sin x}{\sin x} + \frac{1}{\sin x \cos x} = 0$$
$$\frac{\cos x}{\sin x \cos x} - \frac{\sin^2 x}{\sin x \cos x} + \frac{1}{\sin x \cos x} = 0$$
$$\frac{\cos x - \sin^2 x + 1}{\sin x \cos x} = 0$$

Now set the numerator equal to 0, apply the Pythagorean identity, and simplify the equation. This can be factored and the two factors set equal to 0 to solve for x.

$$\begin{aligned}\cos x - \sin^2 x + 1 &= 0\\ \cos x - \left(1 - \cos^2 x\right) + 1 &= 0\\ \cos x + \cos^2 x &= 0\\ \cos x\left(1 + \cos x\right) &= 0\end{aligned}$$

$$\cos x = 0 \qquad\qquad 1 + \cos x = 0$$

$$\cos x = -1$$

$$x = \cos^{-1}(0) = 90,\ 270 \qquad\qquad x = \cos^{-1}(-1) = 180$$

These seem like perfectly nice answers, but they're not. Each of these values for x makes the denominator of the fraction equal to 0 — and you can't have a 0 in the denominator. And, if you look at the original equation, the tangent and secant aren't defined for 90 or 270 degrees, and the cosecant isn't defined for 180 degrees. There's no answer, but that's an answer.

15. Solve for x in $\frac{\cos x}{\sin x} - \frac{\sin x}{\cos x} = 0$ when $0 \le x \le 360$.

Solve It

16. Solve for x in $\frac{1}{\tan x} - \tan x = 0$ when $0 \le x \le 360$.

Solve It

17. Solve for x in $\frac{\sin x + 1}{\cos x} + \cos x = 0$ when $0 \leq x \leq 360$.

Solve It

18. Solve for x in $\cot^2 x - \csc x = 1$ when $0 \leq x \leq 360$.

Solve It

19. Solve for x in $2\tan x - \sin x = 0$ when $0 \leq x \leq 360$.

Solve It

20. Solve for x in $\csc x - 2\cos x = 0$ when $0 \leq x \leq 360$.

Solve It

Using the Quadratic Formula

It's always so nice when a quadratic equation factors and behaves as you're on the road to a solution. This isn't always the case. Luckily, though, there's the quadratic formula to supply the answer — when all else fails. The quadratic formula works on the general quadratic equation $ax^2 + bx + c = 0$, and works just as well on trig functions. The only change is that the x variable will be a function instead.

Q. Use the quadratic formula to solve for x in $\cos^2 x + \cos x - 1 = 0$ when $0 \leq x \leq 360$.

A. $x \approx 38$ degrees or $x \approx 322$ degrees. The values for the quadratic formula in this case are $a = 1$, $b = 1$, and $c = -1$. Substituting into the quadratic formula, $\cos x = \frac{-1 \pm \sqrt{1^2 - 4(1)(-1)}}{2(1)} = \frac{-1 \pm \sqrt{5}}{2}$. The two values you get from this application are $\cos x = \frac{-1 + \sqrt{5}}{2} \approx .618$ or $\cos x = \frac{-1 - \sqrt{5}}{2} \approx -1.618$. The second one will have to be discarded, because the cosine has values between –1 and 1, only. This second equation is impossible. Solving the first for the value of x, $x = \cos^{-1}(.618)$. Look at the chart of values, and you see that the angle closest to having this cosine is 38 degrees. In the fourth quadrant, the angle with a reference angle of 38 degrees is a 322-degree angle. These two angles are the solutions to this equation.

21. Use the quadratic formula to solve $\sin^2 x - \sin x - 1 = 0$ when $0 \leq x \leq 360$.

Solve It

22. Use the quadratic formula to solve $\tan^2 x + 3\tan x - 2 = 0$ when $0 \leq x \leq 360$.

Answers to Problems on Solving Trig Equations

The following are solutions to the practice problems presented earlier in this chapter.

1 Solve for x in $2\sin^2 x - \sin x - 1 = 0$. **$x$ = 90, 210, or 330 degrees.**

First, factor. Then set each factor equal to 0 and solve for x.

$$(2\sin x + 1)(\sin x - 1) = 0$$

$$2\sin x + 1 = 0$$

$$\sin x = -\frac{1}{2}$$

$$x = \sin^{-1}\left(-\frac{1}{2}\right) = 210,\ 330$$

$$\sin x - 1 = 0$$

$$\sin x = 1$$

$$x = \sin^{-1}(1) = 90$$

2 Solve for x in $\cos^2 x - \frac{\sqrt{3}}{2}\cos x = 0$. **$x$ = 30, 90, 270, or 330 degrees.**

First, factor. Then set each factor equal to 0 and solve for x.

$$\cos x\left(\cos x - \frac{\sqrt{3}}{2}\right) = 0$$

$$\cos x = 0$$

$$x = \cos^{-1}(0) = 90,\ 270$$

$$\cos x - \frac{\sqrt{3}}{2} = 0$$

$$\cos x = \frac{\sqrt{3}}{2}$$

$$x = \cos^{-1}\left(\frac{\sqrt{3}}{2}\right)$$

$$= 30,\ 330$$

3 Solve for x in $\tan^2 x - 1 = 0$. **x = 45, 135, 225, or 315 degrees.**

First, factor. Then set each factor equal to 0 and solve for x.

$$(\tan x - 1)(\tan x + 1) = 0$$

$$\tan x - 1 = 0 \qquad \tan x + 1 = 0$$

$$\tan x = 1 \qquad \tan x = -1$$

$$x = \tan^{-1}(1) \qquad x = \tan^{-1}(-1)$$

$$= 45,\ 225 \qquad = 135,\ 315$$

4 Solve for x in $\sec^2 - 4 = 0$. **x = 30, 150, 210, or 330 degrees.**

First, factor. Then set each factor equal to 0 and solve for x.

$$(\sec x - 2)(\sec x + 2) = 0$$

$$\sec x - 2 = 0 \qquad \sec x + 2 = 0$$

$$\sec x = 2 \qquad \sec x = -2$$

$$x = \sec^{-1}(2) \qquad x = \sec^{-1}(-2)$$

$$= 30,\ 330 \qquad = 150,\ 210$$

5 Solve for x in $1 + \cot^2 x + 2\csc x = 0$. **$x$ = 210 or 330 degrees.**

First, replace the first two terms using the Pythagorean identity. Then set each factor equal to 0 and solve for x.

$$
\begin{aligned}
(1+\cot^2 x) + 2\csc x &= 0\\
\csc^2 x + 2\csc x &= 0\\
\csc x(\csc x + 2) &= 0\\
\csc x = 0 \qquad \csc x + 2 &= 0\\
* \qquad \csc x &= -2\\
x &= \csc^{-1}(-2)\\
&= 210,\ 330
\end{aligned}
$$

The first factor didn't yield an answer, because the cosecant is never equal to 0.

6 Solve for x in $\sin 2x - \cos x = 0$. **x = 30, 90, 150, or 270 degrees.**

Replace the double-angle term using the identity. Then factor cos x out of each term. Set each factor equal to 0 and solve for x.

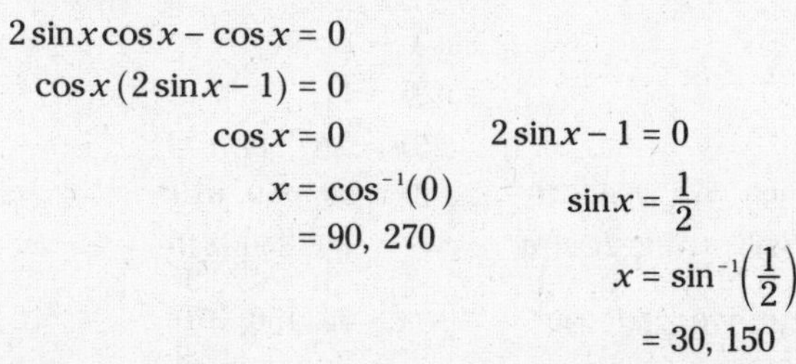

$$
\begin{aligned}
2\sin x\cos x - \cos x &= 0\\
\cos x(2\sin x - 1) &= 0\\
\cos x = 0 \qquad 2\sin x - 1 &= 0\\
x = \cos^{-1}(0) \qquad \sin x &= \frac{1}{2}\\
= 90,\ 270 \qquad x &= \sin^{-1}\left(\frac{1}{2}\right)\\
&= 30,\ 150
\end{aligned}
$$

7 Solve for x in $\sin 2x = -\frac{1}{2}$. **x = 105, 165, 275, 345 degrees.**

First, solve for $2x$. Then find all the angles within *two* revolutions that have that sine. Divide each angle measure by 2.

$$
\begin{aligned}
2x &= \sin^{-1}\left(-\frac{1}{2}\right) = 210,\ 330,\ 570,\ 690\\
x &= \frac{210}{2}, \frac{330}{2}, \frac{570}{2}, \frac{690}{2}\\
&= 105,\ 165,\ 275,\ 345
\end{aligned}
$$

8 Solve for x in $\cos 3x = \frac{\sqrt{2}}{2}$. **$x$ = 15, 105, 135, 225, 255, 345 degrees.**

First, solve for $3x$. Then find all the angles within *three* revolutions that have that cosine. Divide each angle measure by 3.

$$
\begin{aligned}
3x &= 45,\ 315,\ 405,\ 675,\ 765,\ 1035\\
x &= \frac{45}{3}, \frac{315}{3}, \frac{405}{3}, \frac{675}{3}, \frac{765}{3}, \frac{1035}{3}\\
&= 15,\ 105,\ 135,\ 215,\ 255,\ 345
\end{aligned}
$$

9 Solve for x in $\sec^2 2x - \sec 2x = 0$. **x = 0, 180, 360.**

Factor out the sec $2x$. Then set each factor equal to 0. The first factor doesn't yield any solution, because the value of the secant is never 0.

Solve for $2x$. Find all the angles within *two* revolutions that have a secant of 1. Then divide each angle measure by 2.

$$\sec 2x(\sec 2x - 1) = 0$$

$$\sec 2x = 0 \qquad \sec 2x - 1 = 0$$

$$* \qquad \sec 2x = 1$$

$$2x = 0,\ 360,\ 720$$

$$x = \frac{0}{2}, \frac{360}{2}, \frac{720}{2}$$

$$= 0,\ 180,\ 360$$

10 Solve for x in $2\sin^2 3x - 3\sin 3x + 1 = 0$. $\boldsymbol{x}$ **= 10, 30, 50, 130, 150, 170, 250, 270, 290 degrees.**

First, factor and set the two factors equal to 0. Solve for 3*x* in each. Find the angles within *three* revolutions that have that sine. Divide the angle measures by 3.

$$(2\sin 3x - 1)(\sin 3x - 1) = 0$$

$$2\sin 3x - 1 = 0$$

$$\sin 3x = \frac{1}{2}$$

$$3x = \sin^{-1}\left(\frac{1}{2}\right)$$

$$3x = 30,\ 150,\ 390,\ 510,\ 750,\ 870$$

$$x = \frac{30}{3}, \frac{150}{3}, \frac{390}{3}, \frac{510}{3}, \frac{750}{3}, \frac{870}{3}$$

$$x = 10,\ 50,\ 130,\ 170,\ 250,\ 290$$

$$\sin 3x - 1 = 0$$

$$\sin 3x = 1$$

$$3x = \sin^{-1}(1)$$

$$3x = 90,\ 450,\ 810$$

$$x = \frac{90}{3}, \frac{450}{3}, \frac{810}{3}$$

$$x = 30,\ 150,\ 270$$

11 Solve for x in $\csc^3 x = 8$. $\boldsymbol{x}$ **= 30 or 150 degrees.**

Take the cube root of each side and solve for *x*.

$$\sqrt[3]{\csc^3 x} = \sqrt[3]{8}$$

$$\csc x = 2$$

$$x = \csc^{-1}(2) = 30,\ 150$$

12 Solve for x in $\tan^3 x + 1 = 0$. $\boldsymbol{x}$ **= 135 or 315 degrees.**

Subtract 1 from each side. Then take the cube root of each side and solve for *x*.

$$\tan^3 x + 1 = 0$$

$$\tan^3 x = -1$$

$$\sqrt[3]{\tan^3 x} = \sqrt[3]{-1}$$

$$\tan x = -1$$

$$x = \tan^{-1}(-1) = 135,\ 315$$

13 Solve for x in $2\sin x\tan x + \tan x - 2\sqrt{3}\sin x - \sqrt{3} = 0$. $\boldsymbol{x}$ **= 60, 210, 240, or 330 degrees.**

Factor by grouping. Factor tan *x* out of the first two terms and $\sqrt{3}$ out of the second two terms. Then the common factor is 2 sin *x* + 1. Factor it out. Set the two factors equal to 0, and solve for *x*.

$$\tan x(2\sin x+1)-\sqrt{3}(2\sin x+1)=0$$
$$(2\sin x+1)(\tan x-\sqrt{3})=0$$
$$2\sin x+1=0 \qquad \tan x-\sqrt{3}=0$$
$$\sin x=-\frac{1}{2} \qquad \tan x=\sqrt{3}$$
$$x=\sin^{-1}\left(-\frac{1}{2}\right) \qquad x=\tan^{-1}(\sqrt{3})$$
$$=210,\ 330 \qquad =60,\ 240$$

 **14** Solve for x in $\tan x\sec x-2\tan x+\sec x-2=0$. $\boldsymbol{x}$ **= 60, 135, 300, or 315 degrees.**

Factor tan x out of the first two terms and 1 out of the second two. Then factor out the common factor sec $x-2$, set the factors equal to 0, and solve for x.

$$\tan x(\sec x-2)+1(\sec x-2)=0$$
$$(\sec x-2)(\tan x+1)=0$$
$$\sec x-2=0 \qquad \tan x+1=0$$
$$\sec x=2 \qquad \tan x=-1$$
$$x=\sec^{-1}(2) \qquad x=\tan^{-1}(-1)$$
$$=60,\ 300 \qquad =135,\ 315$$

 **15** Solve for x in $\frac{\cos x}{\sin x}-\frac{\sin x}{\cos x}=0$. $\boldsymbol{x}$ **= 45, 135, 225, or 315 degrees.**

First, find a common denominator and subtract the fractions.

$$\frac{\cos x}{\sin x}\cdot\frac{\cos x}{\cos x}-\frac{\sin x}{\cos x}\cdot\frac{\sin x}{\sin x}=0$$
$$\frac{\cos^2 x}{\sin x\cos x}-\frac{\sin^2 x}{\sin x\cos x}=0$$
$$\frac{\cos^2 x-\sin^2 x}{\sin x\cos x}=0$$

The numerator can be replaced using the double-angle identity for the cosine. Then set the numerator equal to 0, solve for $2x$, find all the angles within *two* revolutions with that cosine, and solve for x.

$$\frac{\cos 2x}{\sin x\cos x}=0$$
$$\cos 2x=0$$
$$2x=\cos^{-1}(0)=90,\ 270,\ 450,\ 630$$
$$x=\frac{90}{2},\frac{270}{2},\frac{450}{2},\frac{630}{2}$$
$$=45,\ 135,\ 225,\ 315$$

The denominator of a fraction cannot equal 0. The variable x cannot represent any angle measure that would make one of the functions in the denominator equal to 0.

16 Solve for x in $\frac{1}{\tan x} - \tan x = 0$. **$x$ = 45, 135, 225, or 315 degrees.**

Let tan x be the common denominator. Setting the numerator equal to 0 and factoring, set each factor equal to 0, and solve for x.

$$\frac{1}{\tan x} - \frac{\tan^2 x}{\tan x} = 0$$

$$\frac{1 - \tan^2 x}{\tan x} = 0$$

$$(1 - \tan x)(1 + \tan x) = 0$$

$$1 - \tan x = 0 \qquad 1 + \tan x = 0$$

$$\tan x = 1 \qquad \tan x = -1$$

$$x = \tan^{-1}(1) = 45, 225 \qquad x = \tan^{-1}(-1) = 135, 315$$

17 Solve for x in $\frac{\sin x + 1}{\cos x} + \cos x = 0$. **$x$ = 90 degrees.**

The common denominator is cos x. Add the fractions. Use the Pythagorean identity to change the cosine term. Factor the quadratic in the numerator. Set each factor equal to 0, and solve for x.

$$\frac{\sin x + 1}{\cos x} + \frac{\cos^2 x}{\cos x} = 0$$

$$\frac{\sin x + 1 + \cos^2 x}{\cos x} = 0$$

$$\frac{\sin x + 1 + 1 - \sin^2 x}{\cos x} = 0$$

$$\frac{2 + \sin x - \sin^2 x}{\cos x} = 0$$

$$2 + \sin x - \sin^2 x = 0$$

$$(2 - \sin x)(1 + \sin x) = 0$$

$$2 + \sin x = 0 \qquad 1 - \sin x = 0$$

$$\sin x = 2 \qquad \sin x = -1$$

$$* \qquad x = 270$$

The sine is never greater than 1, so the first choice doesn't work. And the 270 doesn't work, because cos 270=0, and that gives you a 0 in the denominator of the origianl problem.

18 Solve for x in $\cot^2 x - \csc x = 1$. **x = 30, 150, or 270 degrees.**

First, set the equation equal to 0. Rewrite the cotangent term using the ratio identity and the cosecant term using the reciprocal identity. Find a common denominator, and combine the fractions.

$$\cot^2 x - \csc x - 1 = 0$$

$$\frac{\cos^2 x}{\sin^2 x} - \frac{1}{\sin x} - 1 = 0$$

$$\frac{\cos^2 x}{\sin^2 x} - \frac{1}{\sin x} \cdot \frac{\sin x}{\sin x} - 1 \cdot \frac{\sin^2 x}{\sin^2 x} = 0$$

$$\frac{\cos^2 x}{\sin^2 x} - \frac{\sin x}{\sin^2 x} - \frac{\sin^2 x}{\sin^2 x} = 0$$

$$\frac{\cos^2 x - \sin x - \sin^2 x}{\sin^2 x} = 0$$

Replace the cosine term, using the Pythagorean identity. Set the numerator equal to 0, factor it, and set the factors each equal to 0. Solve for x in each.

$$\frac{1 - \sin^2 x - \sin x - \sin^2 x}{\sin^2 x} = 0$$

$$\frac{1 - \sin x - 2\sin^2 x}{\sin^2 x} = 0$$

$$1 - \sin x - 2\sin^2 x = 0$$

$$(1 - 2\sin x)(1 + \sin x) = 0$$

$$1 - 2\sin x = 0$$

$$\sin x = \frac{1}{2}$$

$$x = \sin^{-1}\left(\frac{1}{2}\right) = 30,\ 150$$

$$1 + \sin x = 0$$

$$\sin x = -1$$

$$x = \sin^{-1}(-1) = 270$$

19 Solve for x in $2\tan x - \sin x = 0$. $\boldsymbol{x}$ **= 0 or 180 degrees.**

Rewrite the tangent using the ratio identity. Find a common denominator; then subtract the fractions. Set the numerator equal to 0, factor out $\sin x$, and set the two factors equal to 0. The second factor has no solution, because the cosine is never equal to 2.

$$\frac{2\sin x}{\cos x} - \sin x = 0$$

$$\frac{2\sin x}{\cos x} - \frac{\sin x \cos x}{\cos x} = 0$$

$$\frac{2\sin x - \sin x \cos x}{\cos x} = 0$$

$$2\sin x - \sin x \cos x = 0$$

$$\sin x(2 - \cos x) = 0$$

$$\sin x = 0$$

$$x = \sin^{-1}(0) = 0,\ 180$$

$$2 - \cos x = 0$$

$$\cos x = 2$$

*

20 Solve for x in $\csc x - 2\cos x = 0$. $\boldsymbol{x}$ **= 45 or 225 degrees.**

Rewrite the cosecant using the reciprocal identity. Find a common denominator, and subtract the fractions. Use the double-angle identity for sine to replace the second term in the denominator. Set the denominator equal to 0, and solve for x.

$$\frac{1}{\sin x} - \frac{2\cos x}{1} = 0$$
$$\frac{1}{\sin x} - \frac{2\sin x\cos x}{\sin x} = 0$$
$$\frac{1 - 2\sin x\cos x}{\sin x} = 0$$
$$\frac{1 - \sin 2x}{\sin x} = 0$$
$$1 - \sin 2x = 0$$
$$\sin 2x = 1$$
$$2x = \sin^{-1}(1) = 90,\ 450$$
$$x = \frac{90}{2}, \frac{450}{2}$$
$$= 45,\ 225$$

21 Use the quadratic formula to solve $\sin^2 x - \sin x - 1 = 0$. ***x* is approximately 218 degrees or 322 degrees.**

Use the quadratic formula to solve for sin *x* first. In the formula, $a = 1$, $b = -1$, and $c = -1$.

$$\sin x = \frac{1 \pm \sqrt{1 - 4(1)(-1)}}{2(1)} = \frac{1 \pm \sqrt{5}}{2}$$

Now solve for *x*.

$$\sin x = \frac{1 + \sqrt{5}}{2} \approx 1.618^*$$
$$\sin x = \frac{1 - \sqrt{5}}{2} \approx -.618$$
$$x = \sin^{-1}(-.618) = 218,\ 322$$

The sine is never greater than 1, so the first value doesn't work. The angle measures are given to the nearer degree.

22 Use the quadratic formula to solve $\tan^2 x + 3\tan x - 2 = 0$. ***x* is approximately 29, 106, 209, or 286 degrees.**

Using the quadratic formula, $a = 1$, $b = 3$, and $c = -2$. Solve for tan *x*. Then solve for *x*.

$$\tan x = \frac{-3 \pm \sqrt{9 - 4(1)(-2)}}{2(1)} = \frac{-3 \pm \sqrt{17}}{2}$$

$$\tan x = \frac{-3 + \sqrt{17}}{2} \approx .562$$
$$x = \tan^{-1}(.562) = 29,\ 209$$
$$\tan x = \frac{-3 - \sqrt{17}}{2} \approx -3.562$$
$$x = \tan^{-1}(-3.562) = 106,\ 286$$

Chapter 14

Revisiting the Triangle with New Laws

In This Chapter

- Finding missing parts of triangles with laws of sines and cosines
- Determining areas of triangles
- Applying triangles to practical problems

The right triangle plays a huge role in trigonometry. It's used to establish the trigonometric identities and to solve interesting applications — those where a right angle can be drawn into the picture. How about the other triangles, though? The oblique triangles — those that aren't right (no, they aren't crazy) are just as useful. The oblique triangles come with some laws and formulas that make their application possible.

A ground rule for using these oblique triangles is in the naming of their parts. By convention, the angles are named with capital letters, and the side opposite each angle is named with a corresponding lowercase letter. Look at Figure 14-1 for the naming that will be used in this chapter.

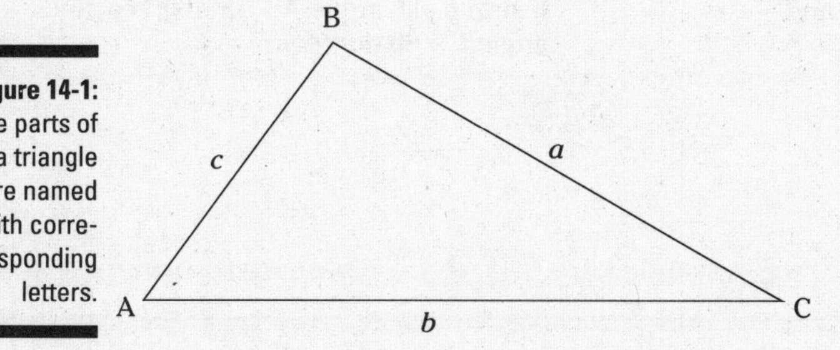

Figure 14-1: The parts of a triangle are named with corresponding letters.

Using the Law of Sines

The Law of Sines says, basically, that the ratios between the sides and sines of the angles opposite those sides of a given triangle are the same. Refer to Figure 14-1 and look at this statement of the Law of Sines. $\frac{\sin A}{a} = \frac{\sin B}{b} = \frac{\sin C}{C}$ or $\frac{a}{\sin A} = \frac{b}{\sin B} = \frac{c}{\sin C}$.

You can take any pair of ratios within the equation, substitute in the three values that you know, and solve for the fourth, or unknown value. This law is used to solve for the missing parts (sides and angles) of a triangle if you know two angles and a side opposite one of them, or two sides and an angle opposite one of them.

Q. Find the three other parts of triangle *ABC*, if you know that angle *A* measures 30 degrees, angle *B* measures 45 degrees, and side *a* measures 10 inches.

A. Angle *C* = 105 degrees; side *b* = 14.14, and side *c* = 5.18. The measure of angle C is determined by just adding the measures of angles A and B and subtracting from 180. 180 – (30 + 45) = 180 – 75 = 105. To find the measure of side *b*, use the Law of Sines with the proportion involving angle *A* and angle *B*. Even though you have exact values for these angle measures, use the table of trig functions and round all answers to three decimal places.

$$\frac{a}{\sin A} = \frac{b}{\sin B}$$

$$\frac{10}{.500} = \frac{b}{.707}$$

$$(.707)\frac{10}{.500} = b$$

$$b = 14.14$$

Now, to find the measure of side *c*, use the proportion involving angles *A* and *C*.

$$\frac{a}{\sin A} = \frac{c}{\sin C}$$

$$\frac{10}{.500} = \frac{c}{.966}$$

$$(.966)\frac{10}{.500} = c$$

$$c = 19.32$$

The sine of 105 degrees is the same as the sine of 75 degrees. (Go to Chapter 8 if you need a review of reference angles.)

1. Find the three other parts of triangle *ABC* if side *a* = 5, angle *A* = 45 degrees, and angle *B* = 70 degrees.

Solve It

2. Find the three other parts of triangle *ABC* if side *b* = 3, angle *A* = 15 degrees, and angle *C* = 60 degrees.

3. A sledding hill has a slope of 40 degrees on one side and a slope of 20 degrees on the other side (see the figure). The 40-degree slope is 100 feet long. How long is the other slope?

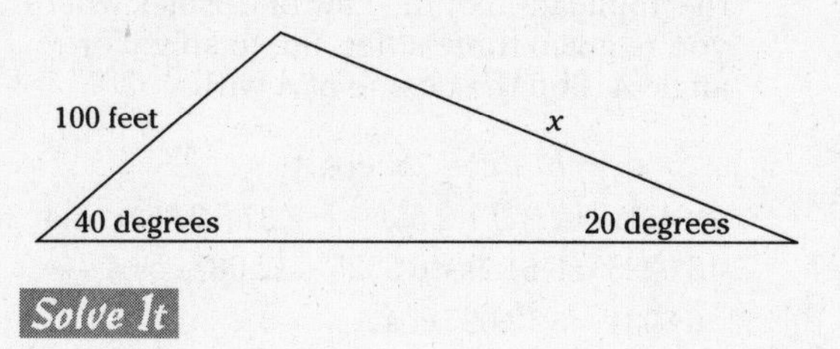

Solve It

4. A telephone pole is leaning to the side, forming an angle of 60 degrees with the ground. A cable is attached to the top of the pole and anchored 40 feet from the pole, on the side away from the direction the pole is leaning (see the figure). The angle that the cable forms with the ground is 45 degrees. How tall is the pole?

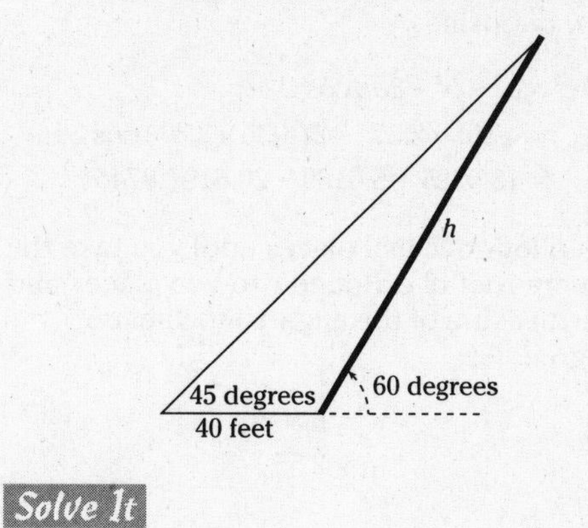

Solve It

Adding the Law of Cosines

The Law of Cosines is used when you have the measures of two sides of a triangle and the angle between them and need to find one or more of the other measures of the triangle. You can't use the Law of Sines unless you have a pair made up of an angle and its opposite side. The Law of Cosines takes care of that shortfall. You can also use this law when you have all three sides.

The Law of Cosines is: $\begin{Bmatrix} a^2 = b^2 + c^2 - 2bc\cos A \\ b^2 = a^2 + c^2 - 2ac\cos B \\ c^2 = a^2 + b^2 - 2ab\cos C \end{Bmatrix}$. It says that the square of the measure of some side of a triangle is equal to the sum of the squares of the other two sides minus two times the measures of those two sides times the cosine of the angle opposite that first side. Whew!

The Law of Cosines has three different versions, but they all mean the same thing — the letters are just different. Essentially, the law solves for the measure of the side opposite the given angle.

Q. Find the measure of the other three parts of the triangle ABC when $a = 4.35$, $b = 2.37$, and angle C = 29 degrees.

A. $c = 2.55$, angle $A = 124$ degrees, angle $B = 27$ degrees. First, solve for side c using the Law of Cosines.

$$c^2 = a^2 + b^2 - 2ab\cos C$$
$$c^2 = 4.35^2 + 2.37^2 - 2(4.35)(2.37)\cos 29$$
$$= 18.9225 + 5.6169 - 20.619(.8746)$$

Keep four decimal places until you take the square root of c. Round c to two places and the measure of the angle to the nearer degree.

$$c^2 = 6.5060$$
$$c = 2.55$$

Now that you have c, it would really be easier to use the Law of Sines to solve for angle A or angle B, but I want to show you the application of the Law of Cosines where you're given three sides. So, to solve for angle A, find the cosine of A with

$$a^2 = b^2 + c^2 - 2bc\cos A$$
$$4.35^2 = 2.37^2 + 2.55^2 - 2(2.37)(2.55)\cos A$$
$$18.9225 = 5.6169 + 6.5025 - 12.087\cos A$$
$$6.8031 = -12.087\cos A$$
$$\cos A = \frac{6.8031}{-12.087} \approx -.5628$$
$$A = \cos^{-1}(-5.628) = 124$$

If angle A is 124 degrees, then angle B is $180 - (29 + 124) = 180 - 153 = 27$ degrees.

5. Use the Law of Cosines to find the length of side c in the triangle ABC if side $a = 4$, side $b = 7$, and angle C is 30 degrees.

Solve It

6. Use the Law of Cosines and the Law of Sines to find the other three parts of the triangle ABC if side $b = 2$, side $c = 5$, and angle A is 150 degrees.

7. Use the Law of Cosines to find the measures of the three angles of the triangle ABC if side $a = 3$, side $b = 4$, and side $c = 6$.

Solve It

8. Find the length of a lake if, from a point in the distance, the north end of the lake is 1,300 meters away, the south end of the lake is 1,000 meters away, and the angle formed by sighting those two points is 45 degrees (see the figure).

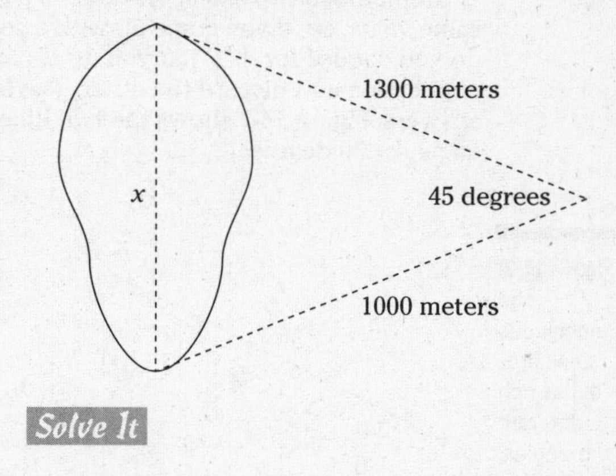

Solve It

9. On a Little League baseball field, the pitcher stands 46 feet from home plate (see the figure). There are 60 feet between the bases. If the lines between the bases form a square, how far is the pitcher from first base? (***Hint:*** The pitcher is not in the middle of the square.)

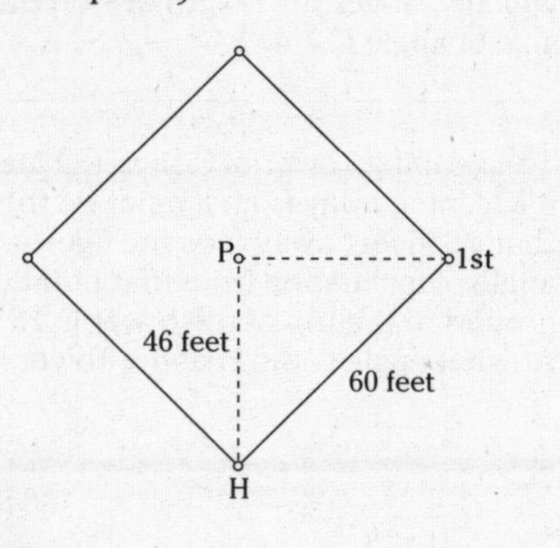

Solve It

10. A solar panel is to be erected on a roof with a pitch of 15 degrees (see the figure). It's to be attached between a place low on the roof to the top of 8-foot-high braces. The braces are attached 10 feet from the lower points. How long must the solar panel be?

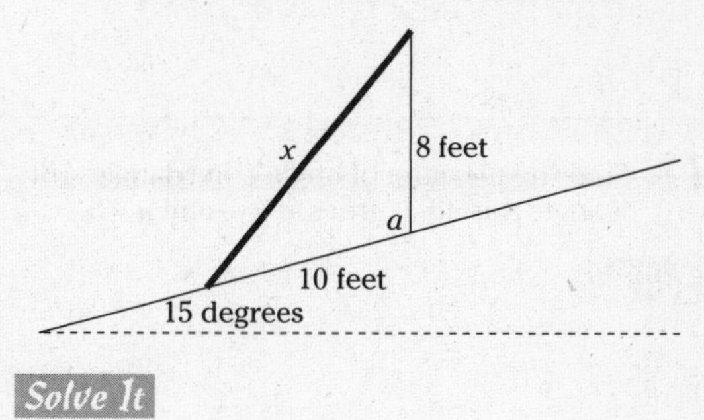

Solve It

Dealing with the Ambiguous Case

When something is *ambiguous,* it can have more than one meaning. The *ambiguous case* in solving triangles occurs when there can be more than one angle that works in the problem. This happens when you have the measures of two sides and an angle not between those two sides. The sine of 30 degrees and the sine of 150 degrees are the same value, so, if you come up with a solution that says that $\sin A = 0.5$, which angle do you choose for *A?* What you do is pick the answer that fits that particular practical application and discard the other. The example here shows how there can be two answers. Figure 14-2 shows the two different triangles that can have $a = 4$, $b = 5$, and angle $A = 30$ degrees.

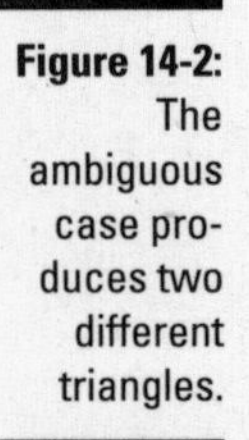

Figure 14-2: The ambiguous case produces two different triangles.

Q. Find the measure of angle B if $a = 4$, $b = 5$, and angle $A = 30$ degrees.

$$\frac{\sin A}{a} = \frac{\sin B}{b}$$

$$\frac{\sin 30}{4} = \frac{\sin B}{5}$$

$$5\left(\frac{.500}{4}\right) = .625 = \sin B$$

$$B = \sin^{-1}(.625) = 39 \text{ or } 141$$

A. 39 degrees or 141 degrees. Using the Law of Sines to solve for the missing measures, first solve for the measure of angle B. As you can see from Figure 14-2, angle B can either be an acute angle or an obtuse angle.

The different choices for the size of angle B will also affect the length of side c and the size of angle C.

11. Find the measure of angle C in triangle ABC if angle A is 45 degrees, $c = 6$, and $a = 5$.

Solve It

12. The Leaning Tower of Pisa is 180 feet tall. It's leaning away from a point on the ground that's 656 feet away (see the figure). The angle of inclination from that point on the ground to the top of the tower is 15 degrees. At what angle is the Leaning Tower leaning?

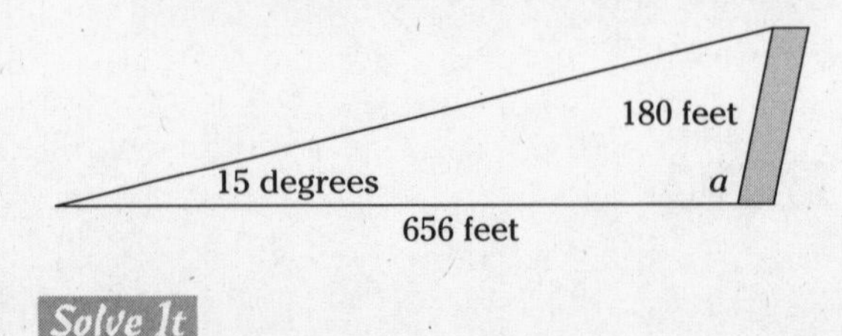

Solve It

Investigating the Law of Tangents

The Law of Tangents isn't used as much as the Law of Sines or the Law of Cosines, but it's unique because it uses the tangent function with the angles that are formed by finding the difference and sum of the angles that occur in the particular application. The Law of Tangents is used in the same situations as the Law of Sines — when you have two angles and a side opposite one of them. Here is the Law of Tangents:

$$\frac{a-b}{a+b}=\frac{\tan\frac{1}{2}(A-B)}{\tan\frac{1}{2}(A+B)}$$

$$\frac{b-c}{b+c}=\frac{\tan\frac{1}{2}(B-C)}{\tan\frac{1}{2}(B+C)}$$

$$\frac{c-a}{c+a}=\frac{\tan\frac{1}{2}(C-A)}{\tan\frac{1}{2}(C+A)}$$

EXAMPLE

Q. Use the Law of Tangents to find the length of side b in triangle ABC when angle A = 86 degrees, side a = 20, and angle B = 40 degrees.

A. b = 13. Substitute the given values into the Law of Tangents. Find the tangents of the two resulting angles, and divide.

$$\frac{a-b}{a+b}=\frac{\tan\frac{1}{2}(A-B)}{\tan\frac{1}{2}(A-B)}$$

$$\frac{20-b}{20+b}=\frac{\tan\frac{1}{2}(86-40)}{\tan\frac{1}{2}(86+40)}$$

$$\frac{20-b}{20+b}=\frac{\tan 23}{\tan 63}=.2163$$

Multiply each side by 20 + b. Then distribute on the right and solve for b.

$$(20+b)\frac{20-b}{20+b}=.2163\,(20+b)$$

$$20-b=4.326+.2163b$$

$$15.674=1.2163b$$

$$b=\frac{15.674}{1.2163}\approx 12.887$$

13. Use the Law of Tangents to find side b when angle A = 60 degrees, angle B = 45 degrees, and side a = 6.

Solve It

14. Use the Law of Tangents to find side a when angle A = 10 degrees, angle B = 100 degrees, and side b = 20.

Solve It

Finding the Area of a Triangle the Traditional Way

The best known formula used to find the area of a triangle is $A = \frac{1}{2}bh$. The only problem with it is that you need to know the measure of a side called the base, *b*, and the measure of a segment that's perpendicular to that side drawn to the vertex opposite the side. This is the height, *h*. What if you can't find the height? That's what the rest of the sections in this chapter are about. This section, though, will deal with the traditional formula, and it will incorporate some properties from trigonometry.

Q. Find the area of a right triangle with a hypotenuse 26 yards long and one leg 24 yards long.

A. $A = 120$ square yards. The two legs of a right triangle are perpendicular to one another, so they form the base and the height. To find the missing leg, use the Pythagorean theorem.

$$24^2 + b^2 = 26^2$$
$$b^2 = 676 - 576 = 100$$
$$b = 10$$

So the two legs are 24 yards and 10 yards. Using the formula for the area,

$$A = \frac{1}{2}bh$$
$$= \frac{1}{2}(10)(24)$$
$$= 120$$

15. Find the area of a 30-60-90 right triangle if the shortest side is 6 inches long.

Solve It

16. Find the area of an isosceles right triangle if the hypotenuse measures $5\sqrt{2}$ feet.

Flying In with Heron's Formula

Heron's formula for finding the area of a triangle requires that you have the measures of the three sides of the triangle. You don't need any perpendicular measure across the inside of the triangle. Let a triangle have sides measuring a, b, and c. Then let s be the *semiperimeter* (half the perimeter — add up the sides and divide by 2). Heron's formula says that the area of the triangle is: $A = \sqrt{s(s-a)(s-b)(s-c)}$.

Q. Find the area of a triangle with sides measuring: $a = 12$, $b = 17$, and $c = 21$.

A. About 101.98 square units. First find the semiperimeter.

$$s = \frac{1}{2}(12 + 17 + 21) = \frac{1}{2}(50) = 25$$

Now use that in Heron's formula:

$$A = \sqrt{25(25-12)(25-17)(25-21)}$$
$$= \sqrt{25(13)(8)(4)}$$
$$= \sqrt{10400} \approx 101.98$$

17. Find the area of a triangle with sides measuring 7, 24, and 25 yards.

Solve It

18. Find the area of a triangle with sides measuring 3, 8, and 10 inches.

Solve It

Finding Area with an Angle Measure

Trigonometry allows you to use another method for finding the area of a triangle. If you know the measure of two sides of the triangle and the angle between those two sides, you can use the formula $A = \frac{1}{2}ab\sin C$, where a and b are the lengths of the sides, and C is the angle between them (the angle opposite side c).

Q. Find the area of the triangle with two of the sides measuring 8 feet and 12 feet with an 80-degree angle between them.

A. About 47.28 square feet. Using the formula

$$A = \frac{1}{2}(8)(12)\sin 80$$
$$= 48(.985)$$
$$= 47.28$$

19. Find the area of the triangle with two sides measuring 5 and 8 feet and an angle of 30 degrees between them.

Solve It

20. Find the area of the triangle with two sides measuring 5 and 8 feet and an angle of 150 degrees between them.

Applying Triangles

The Law of Sines, Law of Cosines, and Law of Tangents allow you to find missing sides and angles in triangles. The various formulas for area provide you with lots of options for finding area, depending on the information that's available. What kinds of applications are there that use these wonderful formulas? You'll see!

Q. Helene stands at the foot of a ladder that reaches to the top of a large boulder. The ladder forms a 30-degree angle with the ground. When Helene walks 50 feet away from the base of the ladder — away from the boulder — the angle of elevation from her feet to the top of the boulder is now 10 degrees. How long is the ladder? (See the figure.)

x

50

10 degrees

30 degrees

A. About 25 feet long. The angle supplementary to the 30-degree angle measures 180 – 30 = 150 degrees. To find the measure of the angle opposite the side measuring 50 feet, 180 – (150 + 10) = 180 – 160 = 20 degrees. Using the Law of Sines,

$$\frac{x}{\sin 10} = \frac{50}{\sin 20}$$

$$\frac{x}{.174} = \frac{50}{.342}$$

$$x = \frac{50}{.342}(.174) \approx 25$$

21. Doug has a triangular yard with sides that measure 100, 200, and 240 yards. He would like to double the size of the yard in terms of square yards. If he doubles the lengths of the sides, will it double the area?

Solve It

22. Tom and Don leave the same place and walk in directions that are 45 degrees apart. Tom walks 12 feet, and Don walks 20 feet. How far apart are they?

Answers to Problems on Triangles

The following are solutions to the practice problems presented earlier in this chapter.

1. Find the three other parts of triangle *ABC* if side $a = 5$, angle $A = 45$ degrees, and angle $B =$ 70 degrees. **Angle $C = 65$ degrees, $b = 6.6$, $c = 6.4$.**

To find the measure of angle C, $180 - (45 + 70) = 180 - 115 = 65$. Using the Law of Sines to find b,

$$\frac{a}{\sin A} = \frac{b}{\sin B}$$
$$\frac{5}{\sin 45} = \frac{b}{\sin 70}$$
$$\frac{5}{.707} = \frac{b}{.940}$$
$$b = \frac{5}{.707}(.940) \approx 6.6$$

Rounded to one decimal place, b is about 6.6. Now, using the Law of Sines to find c,

$$\frac{a}{\sin A} = \frac{c}{\sin C}$$
$$\frac{5}{\sin 45} = \frac{c}{\sin 65}$$
$$\frac{5}{.707} = \frac{c}{.906}$$
$$b = \frac{5}{.707}(.906) \approx 6.4$$

Rounding to one decimal place, c is about 6.4.

2. Find the three other parts of triangle *ABC* if side $b = 3$, angle $A = 15$ degrees, and angle $C =$ 60 degrees. **Angle $B = 105$ degrees, $a = 0.8$, $c = 2.7$.**

To find the measure of angle B, $180 - (15 + 60) = 180 - 75 = 105$. Using the Law of Sines to find a,

$$\frac{a}{\sin A} = \frac{b}{\sin B}$$
$$\frac{a}{\sin 15} = \frac{3}{\sin 105}$$
$$\frac{a}{.259} = \frac{3}{.966}$$
$$a = \frac{3}{.966}(.259) \approx .8$$

Rounded to one decimal place, a is about 0.8. Now, using the Law of Sines to find c,

$$\frac{c}{\sin C} = \frac{b}{\sin B}$$
$$\frac{c}{\sin 60} = \frac{3}{\sin 105}$$
$$\frac{c}{.866} = \frac{3}{.966}$$
$$c = \frac{3}{.966}(.866) \approx 2.7$$

Rounding to one decimal place, c is about 2.7.

3 A sledding hill has a slope of 40 degrees on one side and a slope of 20 degrees on the other side (see the figure). The 40-degree slope is 100 feet long. How long is the other slope? ≈ **188 feet.**

Using the Law of Sines,

$$\frac{x}{\sin 40} = \frac{100}{\sin 20}$$
$$\frac{x}{.643} = \frac{100}{.342}$$
$$x = \frac{100}{.342}(.643) \approx 188$$

4 A telephone pole is leaning to the side, forming an angle of 60 degrees with the ground. A cable is attached to the top of the pole and anchored 40 feet from the pole, on the side away from the direction the pole is leaning (see the figure). The angle that the cable forms with the ground is 45 degrees. How tall is the pole? ≈ **109 feet.**

Using the Law of Sines,

$$\frac{h}{\sin 45} = \frac{40}{\sin 15}$$
$$\frac{h}{.707} = \frac{40}{.259}$$
$$x = \frac{40}{.259}(.707) \approx 109$$

5 Use the Law of Cosines to find the length of side c in the triangle ABC if side $a = 4$, side $b = 7$, and angle C is 30 degrees. ≈ **4 units.**

Using the Law of Cosines,

$$c^2 = a^2 + b^2 - 2ab\cos C$$
$$= 4^2 + 7^2 - 2(4)(7)\cos 30$$
$$= 65 - 48.496 = 16.504$$
$$c \approx 4.06$$

6 Use the Law of Cosines and the Law of Sines to find the other three parts of the triangle ABC if side $b = 2$, side $c = 5$, and angle A is 150 degrees. $\boldsymbol{a = 6.8}$**, angle** $\boldsymbol{B = 8.5}$ **degrees, angle** $\boldsymbol{C =}$ **21.5 degrees.**

First, use the Law of Cosines to solve for the length of side a.

$$a^2 = b^2 + c^2 - 2bc\cos A$$
$$= 2^2 + 5^2 - 2(2)(5)\cos 150$$
$$= 29 - (-17.32) = 46.32$$
$$a \approx 6.8$$

Now, use the Law of Sines to solve for angle B.

$$\frac{\sin A}{a} = \frac{\sin B}{b}$$
$$\frac{\sin 150}{6.8} = \frac{\sin B}{2}$$
$$\frac{.500}{6.8} = \frac{\sin B}{2}$$
$$\sin B = \frac{.500}{6.8}(2) \approx .147$$
$$B = \sin^{-1}(.147) = 8.5 \text{ or } 171.5$$

You have to choose the 8.5-degree angle, because there's already an obtuse angle in this triangle. This means that angle C is 180 – (150 + 8.5) = 180 – 158.5 = 21.5 degrees.

7 Use the Law of Cosines to find the measures of the three angles of the triangle ABC if side $a = 3$, side $b = 4$, and side $c = 6$. **A = 26.4 degrees, B = 36.3 degrees, C = 117.3 degrees.**

Use the Law of Cosines to solve for the cosine of angle A — then find angle A.

$$a^2 = b^2 + c^2 - 2bc\cos A$$
$$3^2 = 4^2 + 6^2 - 2(4)(6)\cos A$$
$$9 = 52 - 48\cos A$$
$$-43 = -48\cos A$$
$$\frac{-43}{-48} = \cos A,\ \cos A \approx .896$$
$$A = \cos^{-1}(.896) = 26.4$$

Using the Law of Cosines to solve for angle B,

$$b^2 = a^2 + b^2 - 2ac\cos B$$
$$4^2 = 3^2 + 6^2 - 2(3)(6)\cos B$$
$$16 = 45 - 36\cos B$$
$$-29 = -36\cos B$$
$$\frac{-29}{-36} = \cos B,\ \cos B \approx .806$$
$$B = \cos^{-1}(.806) = 36.3$$

Then angle C is 180 – (26.4 + 36.3) = 180 – 62.7 = 117.3.

8 Find the length of a lake if, from a point in the distance, the north end of the lake is 1,300 meters away, the south end of the lake is 1,000 meters away, and the angle formed by sighting those two points is 45 degrees (see the figure). **≈ 923 meters.**

Using the Law of Cosines,

$$x^2 = 1300^2 + 1000^2 - 2(1300)(1000)\cos 45$$
$$= 2{,}690{,}000 - 1{,}838{,}478$$
$$= 4{,}528{,}478$$
$$x \approx 923$$

9 On a Little League baseball field, the pitcher stands 46 feet from home plate (see the figure). There are 60 feet between the bases. If the lines between the bases form a square, how far is the pitcher from first base? (***Hint:*** The pitcher is not in the middle of the square.) **≈ 42.6 feet.**

The bases form a square, which has 90-degree angles. The angle between the line from home to the pitcher and the first-base line is 45 degrees. Let x represent the distance from the pitcher to first base. Then, using the Law of Cosines,

$$x^2 = 46^2 + 60^2 - 2(46)(60)\cos 45$$
$$= 5716 - 3903$$
$$= 1813$$
$$x \approx 42.6$$

10 A solar panel is to be erected on a roof with a pitch of 15 degrees (see the figure). It's to be attached between a place low on the roof to the top of 8-foot-high braces. The braces are attached 10 feet from the lower points. How long must the solar panel be? **A little over 14 feet.**

Angle A is 15 degrees greater than a 90-degree angle, or 105 degrees. Using the Law of Cosines,

$$\begin{aligned} x^2 &= 8^2 + 10^2 - 2(8)(10)\cos 105 \\ &= 164 - (-41.4) \\ &= 205.4 \\ x &\approx 14.3 \end{aligned}$$

11 Find the measure of angle C in triangle ABC if angle A is 45 degrees, $c = 6$, and $a = 5$. **Either 58 degrees.**

Using the Law of Sines,

$$\begin{aligned} \frac{\sin A}{a} &= \frac{\sin C}{c} \\ \frac{\sin 45}{5} &= \frac{\sin C}{6} \\ \frac{.707}{5} &= \frac{\sin C}{6} \\ \sin C &= \frac{.707}{5}(6) = .848 \\ C &= \sin^{-1}(.848) = 58 \end{aligned}$$

12 The Leaning Tower of Pisa is 180 feet tall. It's leaning away from a point on the ground that's 656 feet away (see the figure). The angle of inclination from that point on the ground to the top of the tower is 15 degrees. At what angle is the Leaning Tower leaning? **A little more than 4 degrees off vertical.**

Let a represent the angle you want and let x represent the angle at the top of the triangle, near the top of the tower. Then, using the Law of Sines,

$$\begin{aligned} \frac{\sin 15}{180} &= \frac{\sin x}{656} \\ \frac{.259}{180} &= \frac{\sin x}{656} \\ \sin x &= \frac{.259}{180}(656) = .944 \\ x &= \sin^{-1}(.944) = 70.7 \text{ or } 109.3 \end{aligned}$$

x can be either 70.7 degrees or 109.3 degrees. From the figure, I chose 70.7 degrees. To find angle A, $180 - (70.7 + 15) = 180 - 85.7 = 94.3$. This is $94.3 - 90 = 4.3$ degrees over from the vertical.

13 Use the Law of Tangents to find side b when angle $A = 60$ degrees, angle $B = 45$ degrees, and side $a = 6$. **$b = 4.9$.**

Using the Law of Tangents,

$$\begin{aligned} \frac{a-b}{a+b} &= \frac{\tan\frac{1}{2}(A-B)}{\tan\frac{1}{2}(A+B)} \\ \frac{6-b}{6+b} &= \frac{\tan\frac{1}{2}(60-45)}{\tan\frac{1}{2}(60+45)} \\ \frac{6-b}{6+b} &= \frac{\tan\frac{1}{2}(15)}{\tan\frac{1}{2}(105)} = \frac{.132}{1.303} = .101 \\ 6-b &= .101(6+b) \\ 6-b &= .606 + .101b \\ 5.394 &= 1.101b \\ b &= \frac{5.394}{1.101} \approx 4.9 \end{aligned}$$

14 Use the Law of Tangents to find side a when angle A = 10 degrees, angle B = 100 degrees, and side b = 20. $\boldsymbol{a}$ **= 3.5.**

Using the Law of Tangents,

$$\frac{a-20}{a+20}=\frac{\tan\frac{1}{2}(10-100)}{\tan\frac{1}{2}(10+100)}$$

$$\frac{a-20}{a+20}=\frac{\tan\frac{1}{2}(-90)}{\tan\frac{1}{2}(110)}=\frac{-1}{1.428}=-.700$$

$$a-20=-.700\,(a+20)$$
$$a-20=-.7a-14$$
$$1.7a=6$$
$$a=\frac{6}{1.7}\approx 3.5$$

15 Find the area of a 30-60-90 right triangle if the shortest side is 6 inches long. ≈ **31.2 square inches.**

If the shortest side is 6 inches long, the other leg is $\sqrt{3}$ times that or $6\sqrt{3}$ inches long. The area is $A=\frac{1}{2}(6)(6\sqrt{3})=18\sqrt{3}\approx 31.2$.

16 Find the area of an isosceles right triangle if the hypotenuse measures $5\sqrt{2}$ feet. **12.5 square feet.**

The length of the hypotenuse of an isosceles right triangle is $\sqrt{2}$ times the leg, so the legs measure 5 feet. The area is $\frac{1}{2}(5)(5)=12.5$.

17 Find the area of a triangle with sides measuring 7, 24, and 25 yards. **84 square yards.**

The semiperimeter is $s=\frac{1}{2}(7+24+25)=28$. The area is

$$A=\sqrt{28\,(28-7)(28-24)(28-25)}$$
$$=\sqrt{28\,(21)(4)(3)}$$
$$=\sqrt{7056}$$
$$=84$$

18 Find the area of a triangle with sides measuring 3, 8, and 10 inches. ≈ **9.9 square inches.**

The semiperimeter is $s=\frac{1}{2}(3+8+10)=10.5$. The area is

$$A=\sqrt{10.5\,(10.5-3)(10.5-8)(10.5-10)}$$
$$=\sqrt{10.5\,(7.5)(2.5)(.5)}$$
$$=\sqrt{98.4375}$$
$$\approx 9.9$$

19 Find the area of the triangle with two sides measuring 5 and 8 feet and an angle of 30 degrees between them. **10 square feet.**

The area is $A=\frac{1}{2}(5)(8)\sin 30=20\,(.5)=10$.

20 Find the area of the triangle with two sides measuring 5 and 8 feet and an angle of 150 degrees between them. **10 square feet.**

The area is $A=\frac{1}{2}(5)(8)\sin 150=20\,(.5)=10$. Notice how this is the same area as in problem 21. The sine of 30 degrees and 150 degrees is the same. The following figure illustrates the two triangles. They have the same area.

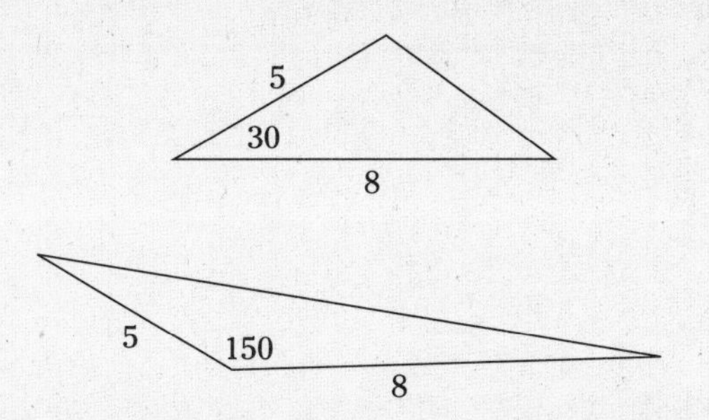

21 Doug has a triangular yard with sides that measure 100, 200, and 240 yards. He would like to double the size of the yard in terms of square yards. If he doubles the lengths of the sides, will it double the area? **No.**

Doug will get much more than twice the area by doubling the sides. This is the effect of multiplying the lengths and using square measures. The area of the original yard is found with Heron's formula. The semiperimeter is $s = \frac{1}{2}(100 + 200 + 240) = 270$. The area is

$$\begin{aligned} A &= \sqrt{270(270-100)(270-200)(270-240)} \\ &\approx 9818 \end{aligned}$$

Doubling the sides, the new semiperimeter is $s = \frac{1}{2}(200 + 400 + 480) = 540$ and the new area is

$$\begin{aligned} A &= \sqrt{540(540-200)(540-400)(540-480)} \\ &\approx 39271 \end{aligned}$$

22 Tom and Don leave the same place and walk in directions that are 45 degrees apart. Tom walks 12 feet, and Don walks 20 feet. How far apart are they? **≈ 14.3 feet.**

Using the Law of Cosines,

$$\begin{aligned} x^2 &= 12^2 + 20^2 - 2(12)(20)\cos 45 \\ &= 544 - 339.41 \\ &= 204.59 \\ x &\approx 14.3 \end{aligned}$$

Part IV
Graphing the Trigonometric Functions

"He was there a minute ago. I just stepped out as he was running some trig equations in order to calculate the area of the Bermuda Triangle."

In this part . . .

This part could also be called the EKG: Extra Kindly Graphs. Graphing the trig functions takes special care at the onset, but then they're so very cooperative. The graphs repeat themselves, so you always know what's going on with them. I show you how to find the points of reference for each graph, and then you see how to sketch in the rest of the graph from them. I introduce you to the world of asymptotes — those lines that aren't really there. You'll see how this all fits together to draw an accurate graph.

Chapter 15

Graphing Sine and Cosine

In This Chapter

- Plotting graphs of sine and cosine by special points
- Incorporating changes in the graphs
- Using graphs of sine and cosine in real life

The graphs of the sine and cosine functions are used extensively to model what goes on every day, every week, and every year. These functions are *periodic*. They repeat their values over and over, in predictable periods of time. Seasonal sales, temperatures, moon phases, and other phenomena are illustrated by some version of the sine or cosine function.

The graphs of the sine and cosine functions look like waves moving across the page (see Figure 15-1). The details involved in these graphs are covered in the sections in this chapter.

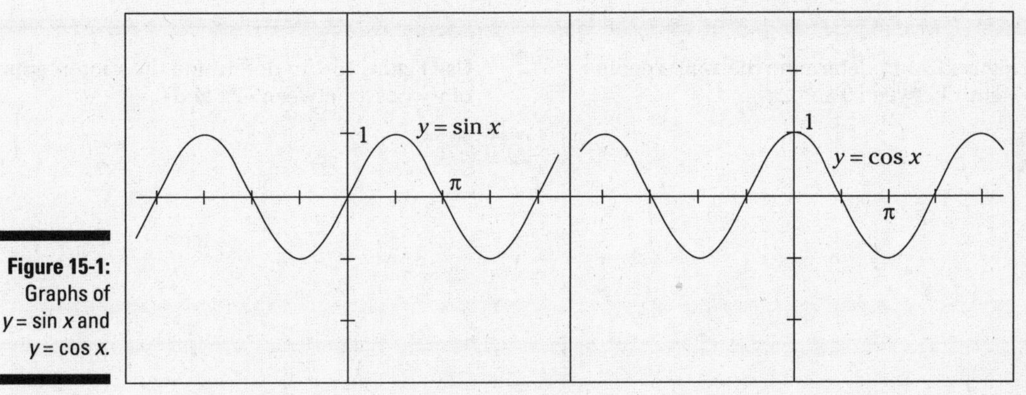

Figure 15-1: Graphs of $y = \sin x$ and $y = \cos x$.

The graphs in Figure 15-1 are shown on the coordinate system, with each tick mark on the x-axis representing $\frac{\pi}{2}$ units. That's about 1.57 units in the real-number system, and it's equal to 90 degrees. Graphs of the trig functions in this chapter use these multiples of π on the horizontal axis. This way, both the x and y axes are using the same number system — the real numbers.

Determining Intercepts and Extreme Values

The x-intercepts of a function are the points at which the function crosses the x-axis. The coordinates of these points look like (1,0) or (–2,0). The y-coordinate is always 0. The y-intercept is the place where the function crosses the y-axis. There's only one y-intercept for any function (that's part of its definition). The coordinate always has an x-value that's 0. Graphing the trig functions requires that you know where the intercepts are.

Extreme values of the sine and cosine functions are where they're the highest or lowest. The sine and cosine functions are periodic, so these highs and lows will occur over and over in a regular, predictable pattern. The functions $y = \sin x$ and $y = \cos x$ have extreme values when $y = 1$ or $y = -1$.

Q. What are the x-intercepts and points where there are extreme values of the function $y = \sin x$ between -2π and 0? Refer to Figure 15-1.

A. The x-intercepts shown in the figure are $(-2\pi, 0)$, $(-\pi, 0)$, and $(0, 0)$. The extreme values are at $\left(-\frac{\pi}{2}, -1\right)$ and $\left(-\frac{3\pi}{2}, 1\right)$.

1. Use Figure 15-1 to determine the x-intercepts of $y = \sin x$ between 0 and 2π.

Solve It

2. Use Figure 15-1 to determine the x-intercepts of $y = \cos x$ between -2π and 0.

Solve It

3. Use Figure 15-1 to determine the extreme points of $y = \sin x$ between 0 and 2π.

Solve It

4. Use Figure 15-1 to determine the extreme points of $y = \cos x$ between 0 and 2π.

Graphing the Basic Sine and Cosine Curves

The graphs of the sine and cosine curves are already done for you in Figure 5-1, but the technique outlined here is used when graphing any variation on either of these functions. Use the following procedure when graphing these functions:

1. **Draw x and y axes, marking the x axis with tick marks that are a multiple of π and the y axis with integers to fit the problem.**
2. **Determine all x-intercepts by solving an equation such as $x = \sin^{-1}(0)$ or $x = \cos^{-1}(0)$ for all x-values in the interval you're graphing.**
3. **Graph the intercepts.**
4. **Determine the y-intercept by letting $x = 0$ and solving for y in the equation.**
5. **Graph the y-intercept.**
6. **Determine where the extreme values are by solving an equation such as $x = \sin^{-1}(\pm 1)$ or $x = \cos^{-1}(\pm 1)$ for all x-values in the interval you're graphing.**
7. **Graph those values.**
8. **Sketch in the curve.**

Q. Graph $y = \sin x$ between $x = 2\pi$ and $x = 4\pi$.

A. To find the x-intercepts, solve $x = \sin^{-1}(0)$ to get $x = 2\pi$ or 3π or 4π. There's no y-intercept here, because the y-axis isn't in that interval. The function is equal to 1 at $x = \frac{5\pi}{2}$ and -1 at $x = \frac{7\pi}{2}$. The graph is shown in the figure.

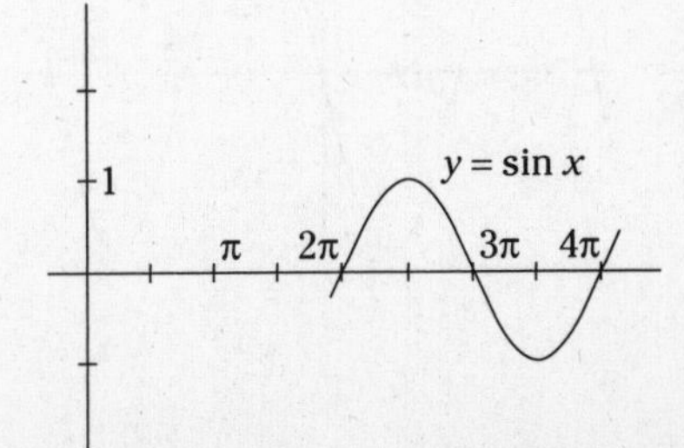

5. Sketch the graph of $y = \sin x$ between -2π and 3π.

Solve It

6. Sketch the graph of $y = \cos x$ between 0 and 4π.

Solve It

Changing the Amplitude

The *amplitude* of the sine and cosine function is a number telling how much it deviates from the average value or middle value. In the graphs of $y = \sin x$ and $y = \cos x$, the amplitude is 1, because each of these curves has a maximum y value of 1 and a minimum y value of -1. The deviation from the average or middle value (0) is 1 in either direction. You can tell what the amplitude is by looking at the coefficient (multiplier) of $\sin x$ or $\cos x$ in the equation. The amplitude of $y = 5\sin x$ is 5; the graph goes to a high of 5 and a low of -5.

Q. Sketch the graph of $y = 2\sin x$.

A. Notice that, where the extreme values used to be 1 or -1, they're now 2 or -2. The graph is shown in the figure.

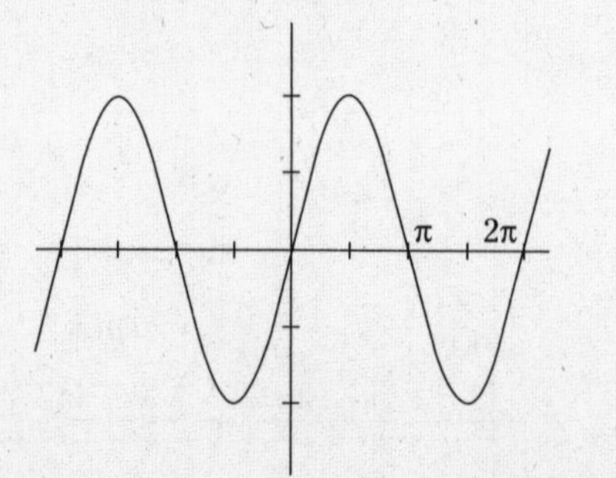

Q. Sketch the graph of $y = -\frac{1}{2}\cos x$.

A. The amplitude is ½. The negative in front of the coefficient doesn't really affect the amplitude. Just think of the amplitude as being the absolute value of that number. The negative part just changes where the maximum and minimum values occur. The graph is shown in the figure.

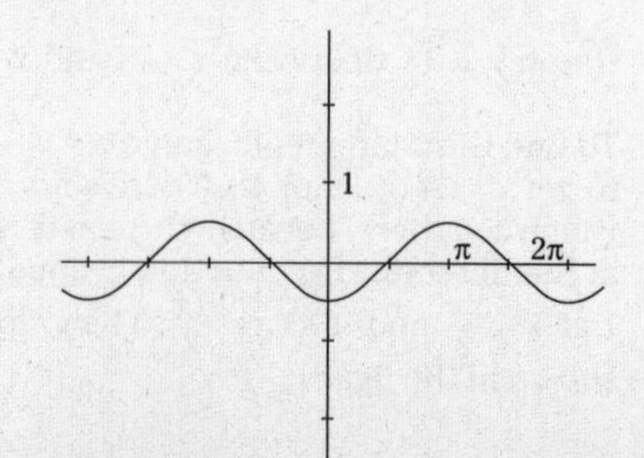

7. Sketch the graph of $y = 3\sin x$.

Solve It

8. Sketch the graph of $y = 4\cos x$.

Solve It

9. Sketch the graph of $y = -2\cos x$.

Solve It

10. Sketch the graph of $y = \frac{1}{3}\sin x$.

Solve It

15. Identify the amplitude, period, shift left or right, and shift up or down, and then sketch a graph of the function $y = 3\sin 2x$.

16. Identify the amplitude, period, shift left or right, and shift up or down, and then sketch a graph of the function $y = \sin\left(x + \frac{\pi}{2}\right)$.

Solve It

17. Identify the amplitude, period, shift left or right, and shift up or down, and then sketch a graph of the function $y = -2\cos x + 3$.

Solve It

18. Identify the amplitude, period, shift left or right, and shift up or down, and then sketch a graph of the function $y = -\frac{1}{3}\sin 3x$.

19. Identify the amplitude, period, shift left or right, and shift up or down, and then sketch a graph of the function $y = \sin\left(\frac{1}{2}x\right) - 1$.

Solve It

20. Identify the amplitude, period, shift left or right, and shift up or down, and then sketch a graph of the function $y = 5\cos 2x + 2$.

Solve It

Applying the Sine and Cosine Curves to Life

I keep talking about how the sine and cosine curves can model what goes on in real life. Now's the time to come through and show you an example.

Q. The temperature in a refrigerator doesn't stay the same — even when you keep the door closed. The compressor comes on to cool it down, whenever the temperature gets too high inside. Here's a sine curve that models the temperature in a particular refrigerator: $y = 2.6 \sin(0.1x - 0.5) + 35$, where the temperature is in degrees Fahrenheit, and x is the number of minutes. What is the average temperature? How high does it rise, and how low does the temperature fall? How often does the compressor work?

A. First, look at the graph of this model, shown in the figure. This graph shows you how the temperature stays within a certain range — between 37.6 degrees and 32.4 degrees — the high and low. How do you get that range? The average temperature is 35 degrees. So you add and subtract the amplitude, 2.6, from that average to get the high and the low. The amount of time it takes to go through a complete cycle is its period. And the period is found by dividing 2π by 0.1, which comes out to be about 62.8 minutes — a little over one hour. The shift to the right is 5 minutes (divide 0.5 by 0.1), which means that the cycle starts 5 minutes after the beginning of the hour.

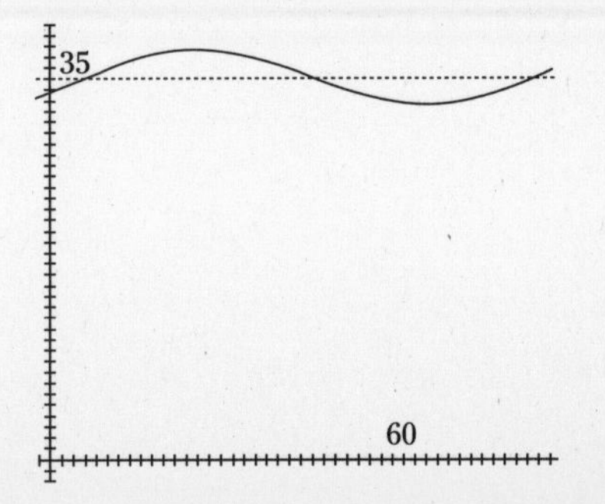

7 Sketch the graph of $y = 3\sin x$. **See the following figure.**

The amplitude is 3, so the graph goes from a low of –3 to a high of 3.

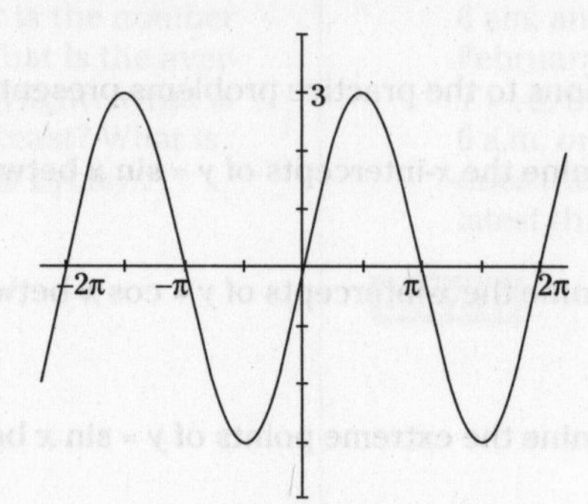

 Sketch the graph of $y = 4\cos x$. **See the following figure.**

The amplitude is 4, so there's a low of –4 and a high of 4.

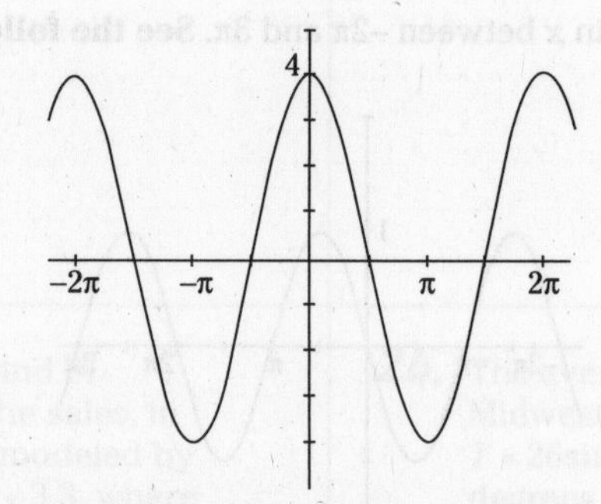

9 Sketch the graph of $y = -2\cos x$. **See the following figure.**

The –2 flips things over. The high is still 2 and the low is still –2, but the highs and lows are in the opposite positions as they are when the coefficient is positive.

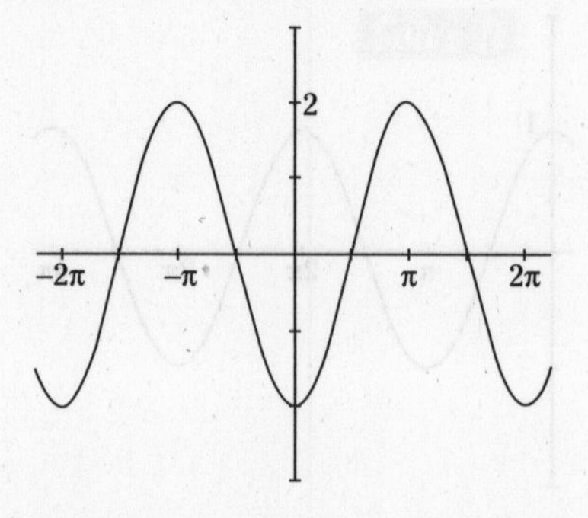

10 Sketch the graph of $y = \frac{1}{3}\sin x$. **See the following figure.**

This fractional multiplier makes the graph shrink.

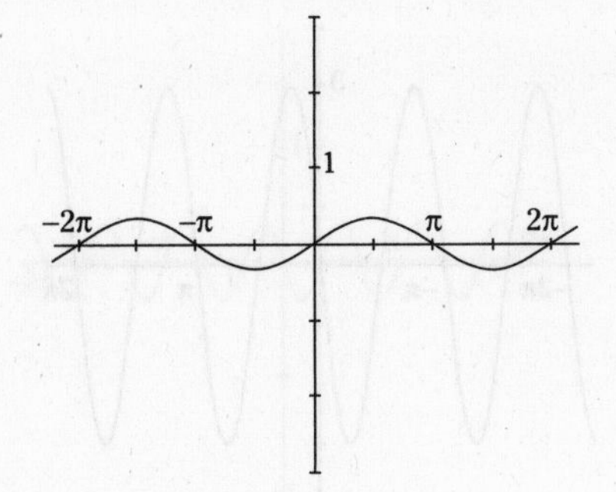

11 Find the period of $y = \sin 4x$. **The period is $\frac{\pi}{2}$.**

You get this by dividing 2π by 4.

12 Find the period of $y = \cos\frac{1}{2}x$. **The period is 4π.**

You get this when you divide 2π by ½.

13 Find the period of $y = -3\cos 3x$, and graph the curve. **The period is $\frac{2\pi}{3}$; see the following figure.**

You get three complete curves where there's usually only one. The amplitude is 3, so it's going to go from −3 to 3.

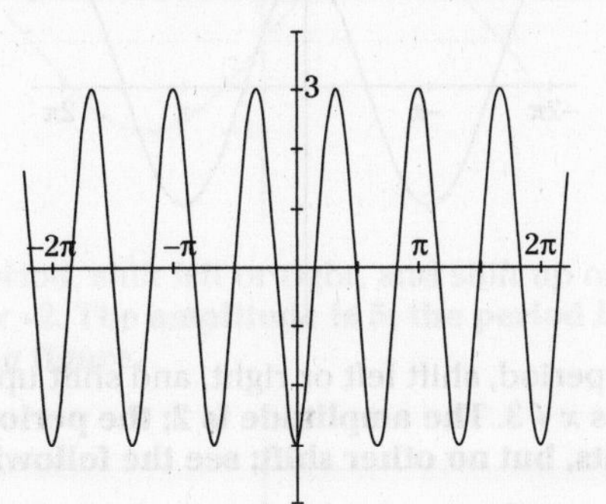

14 Find the period of $y = -\frac{1}{4}\sin\frac{1}{2}x$, and graph the curve. **The period is 4π; see the following figure.**

It takes much longer to go through a complete cycle. The amplitude is pretty small, too.

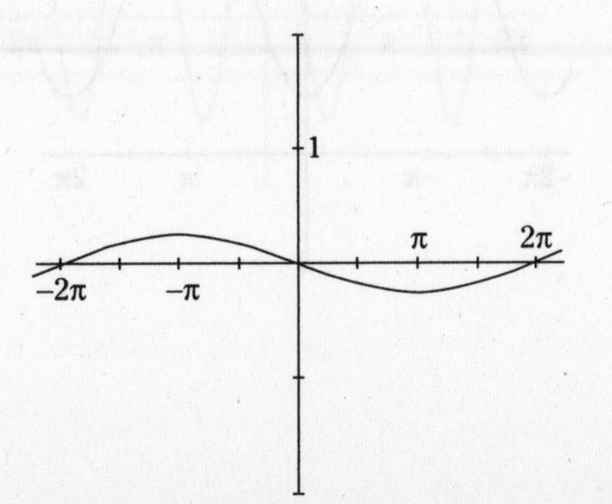

21 The number of hours of daylight in a particular city can be modeled by $H = 2.5\sin(0.02x - 1.28) + 12$, where H is the number of hours and x is the number of days since January 1. What is the average number of hours of daylight? What is the most daylight and the least? What is the period (how long is this cycle)? **The average number of hours of daylight is 12. The most daylight is 14.5 hours, and the least daylight is 9.5 hours. The period is about 314 days.**

The average number of hours is 12; that comes from the 12 at the end of the equation. The most daylight is $12 + 2.5 = 14.5$, and the least amount is $12 - 2.5 = 9.5$. You get the period by dividing 2π by 0.02, to arrive at about 314.

22 The time that the sun rises can be modeled by $T = 40\sin(0.524x + 1.53) + 1.75$ where T is the number of minutes before or after 6 am, and x is the month (January = 1, February = 2, and so on). To use this, if $T = -3$, that corresponds to 3 minutes before 6 a.m. or 5:57 a.m. On average, what time does the sun rise? What is the earliest and latest that it rises? How long is this cycle? **On average, the sun rises at about 6:02 a.m. The earliest it rises is 5:22 a.m., and the latest it rises is 6:42 a.m. The cycle is 12 months.**

You get the average sunrise by adding 1.75 minutes to 6 a.m., to get about 6:02 a.m. The earliest it rises is 40 minutes before the average, or 5:22 a.m. The latest is 40 minutes after the average, or 6:42 a.m. The cycle is determined by dividing 2π by 0.524.

23 The sale of a particular brand of soccer shoe is seasonal. The sales, in millions of dollars, can be modeled by $M = 2.4\sin(0.524x - 0.942) + 3.3$, where x is the month of the year. What are the highest and lowest levels of sales? What is the average? **The highest level of sale is $5.7 million, and the lowest level is $900,000 (or $0.9 million). The average is $3.3 million.**

To get the amount of the highest level, add $3.3 + 2.4$. This is the average value plus the amplitude. The average value is that constant amount at the end of the equation.

24 The average daily temperature in a Midwestern city can be found with $T = 26\sin(0.017x - 1.9) + 48$, where T is in degrees Fahrenheit, and x is the day of the year (January 1 = 1, February = 2, and so on). What is the highest average temperature? What is the lowest average temperature? What is the length of this cycle? **The highest average temperature is 74 degrees, and the lowest is 22 degrees. The length of this cycle is about 370 days.**

You get the high by adding $48 + 26$, the average temperature of 48 plus the amplitude of 26; the lowest value is $48 - 26$. Then, to get the length of the cycle, divide 2π by 0.017.

Chapter 16

Graphing Tangent and Cotangent

In This Chapter

- Drawing in vertical asymptotes
- Sketching tangent and cotangent

The graphs of the tangent and cotangent functions are very similar. They look like flattened-out S-shaped curves. The main differences are that one graph moves upward from left to right, while the other moves downward, and that they have different intercepts. Graphing these curves just takes knowing how to recognize these two characteristics and then fitting the graphs to them.

Establishing Vertical Asymptotes

The tangent and cotangent functions are not defined for all real numbers. Each can be written as a *ratio* (a fraction) of the sine and cosine. Because of this, when the sine or cosine is equal to zero and in the denominator of that ratio, an undefined value occurs. The way this is shown in the graph of the function is to sketch in a *vertical asymptote.* A vertical asymptote isn't a part of any graph. It's a line that's drawn in with dots or dashes to show where it is but indicate that it's just there for help in drawing. Asymptotes, in general, help with the shape of a curve when you're drawing it in.

The vertical asymptotes for $y = \tan x$ occur when the cosine is equal to 0. That occurs when $x = \frac{k\pi}{2}$; k is some odd integer. This means that they occur for any odd multiple of $\frac{\pi}{2}$. The vertical asymptotes for $y = \cot x$ occur when the sine is equal to 0. That happens when $x = k\pi$, or any multiple of π. Figure 16-1 shows what the asymptotes look like.

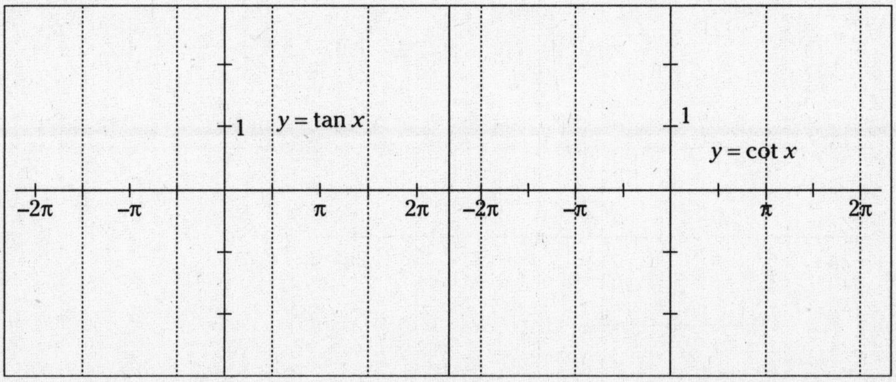

Figure 16-1: Asymptotes for $y = \tan x$ (left) and $y = \cot x$ (right).

Q. Name the vertical asymptotes for $y = \cot x$ between $x = 0$ and $x = 2\pi$.

A. The vertical asymptotes are: $x = 0$, $x = \pi$, and $x = 2\pi$. The sine is 0 for each of these values.

1. Name all the vertical asymptotes for $y = \tan x$ between $x = -3\pi$ and $x = 3\pi$.

Solve It

2. Name all the vertical asymptotes for $y = \cot x$ between $x = -3\pi$ and $x = 3\pi$.

Graphing Tangent and Cotangent

The graphs of the tangent and cotangent function are essentially flattened-out curves that fit between the vertical asymptotes. The graph of the tangent function moves upward as you move from left to right. It crosses the *x*-axis halfway between the asymptotes. The same curve keeps repeating over and over. The graph of the cotangent function moves downward as you move from left to right. It also crosses the *x*-axis halfway between its asymptotes. In Figures 16-2 and 16-3, I've drawn in one piece of the graph of each. You get to draw in more of the graphs in the exercises in this section.

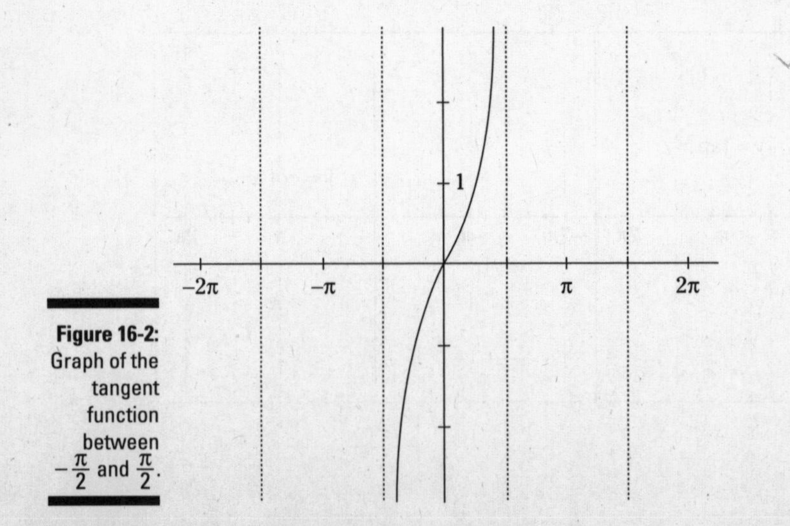

Figure 16-2: Graph of the tangent function between $-\frac{\pi}{2}$ and $\frac{\pi}{2}$.

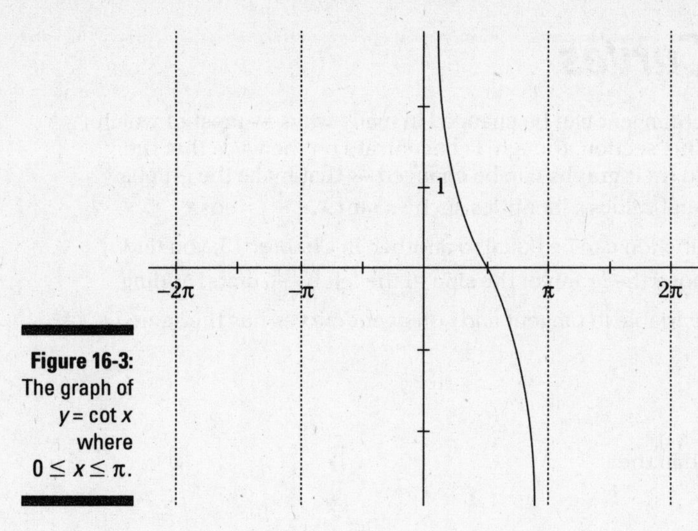

Figure 16-3: The graph of $y = \cot x$ where $0 \leq x \leq \pi$.

Q. **Draw the asymptotes and graph of $y = \cot x$ between $x = 0$ and $x = \pi$.**

A. **Refer to Figure 16-3 for the asymptotes and graph.**

3. Sketch the graph of $y = \tan x$ from $x = -3\pi$ to $x = 3\pi$.

Solve It

4. Sketch the graph of $y = \cot x$ from $x = -3\pi$ to $x = 3\pi$.

Altering the Basic Curves

The graphs of the tangent and cotangent can be changed in many ways — most of which are covered in Chapter 18. For this section, though, I concentrate on the ways that the functions can be changed — and their graphs can be changed — that make the graphs appear to be the same (have equal values). Identities such as $\sin\left(x+\frac{\pi}{2}\right)=\cos x$ confirm that the graph of one function can be equal to another. In Chapter 15, you find that $y=\sin\left(x+\frac{\pi}{2}\right)$ means to move the graph of the sine to the left by $\frac{\pi}{2}$ units. Adding or subtracting values from the variable in tangent and cotangent curves has the same effect.

Q. Graph $y = \tan(x + \pi)$. What function has the same graph?

A. It's the same as $y = \tan x$. Moving the graph π units to the left results in the asymptotes falling onto one another and the intercepts lying on top of one another. It's the same graph. You can see this in the figure.

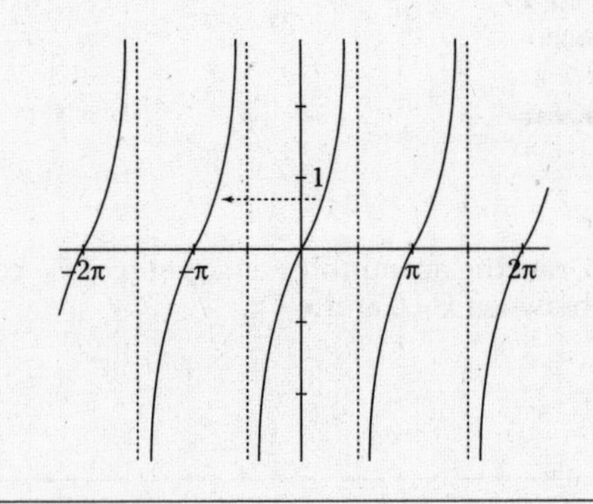

5. Sketch the graph of $y = \tan(x - \pi)$. What function has the same graph?

Solve It

6. Sketch the graph of $y = \cot(x + 2\pi)$. What function has the same graph?

Answers to Problems on Graphing Tangent and Cotangent

The following are solutions to the practice problems presented earlier in this chapter.

1 Name all the vertical asymptotes for $y = \tan x$ between $x = -3\pi$ and $x = 3\pi$. $\boldsymbol{x = -\frac{5\pi}{2}, x = -\frac{3\pi}{2}, x = -\frac{\pi}{2}, x = \frac{\pi}{2}, x = \frac{3\pi}{2}, x = \frac{5\pi}{2}}$.

The asymptotes are all odd multiples of $\frac{\pi}{2}$.

2 Name all the vertical asymptotes for $y = \cot x$ between $x = -3\pi$ and $x = 3\pi$. $\boldsymbol{x = -3\pi, x = -2\pi, x = -\pi, x = 0, x = \pi, x = 2\pi, x = 3\pi}$.

The asymptotes are all multiples of π.

3 Sketch the graph of $y = \tan x$ from $x = -3\pi$ to $x = 3\pi$. **See the following figure.**

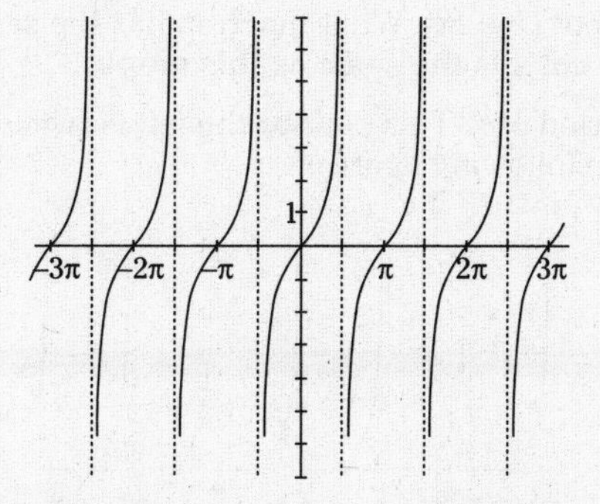

4 Sketch the graph of $y = \cot x$ from $x = -3\pi$ to $x = 3\pi$. **See the following figure.**

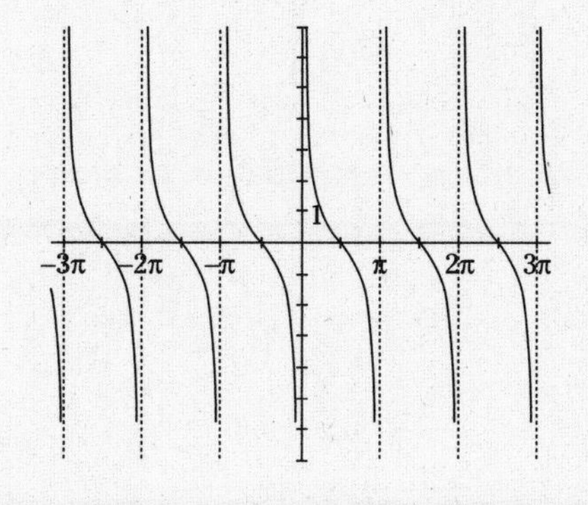

5 Sketch the graph of $y = \tan(x - \pi)$. What function has the same graph? **See the following figure; the graph of $y = \tan x$ is the same as this graph.**

The tangent has a period of π, and this shift to the right is exactly π in length. The graph of one falls on top of the other (see the following figure).

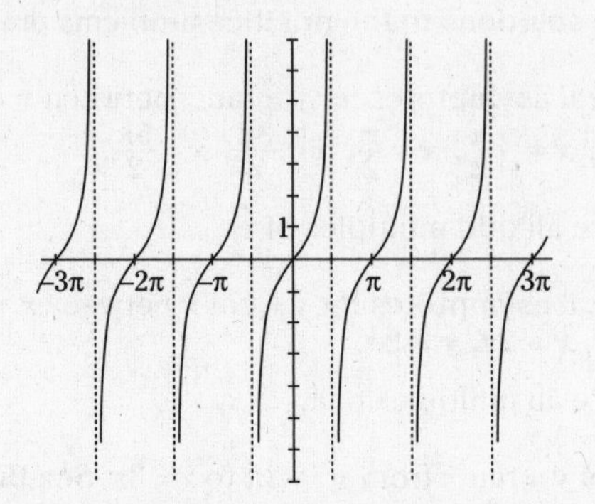

6 Sketch the graph of $y = \cot(x + 2\pi)$. What function has the same graph? **See the following figure; the graph of $y = \cot x$ is the same as this graph.**

The cotangent has a period of π. This shift to the left is twice that, so the graph of one falls on top of the other (see the following figure).

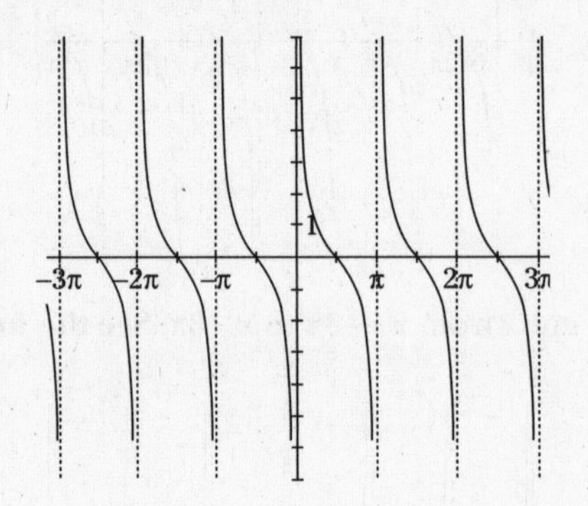

Chapter 17

Graphing Cosecant, Secant, and Inverse Trig Functions

In This Chapter

- Graphing secant and cosecant
- Relating functions to reciprocals
- Graphing inverse trig functions

The secant and cosecant functions are reciprocals of the cosine and sine functions, respectively. Their graphs don't look anything like their reciprocals, but the graphs of the sine and cosine are very helpful when doing the graphing of the cosecant and secant. There are asymptotes to consider in these graphs, as with the tangent and cotangent. But there are no x-intercepts to help with the graphing.

Determining the Vertical Asymptotes

Vertical asymptotes (sort of "ghost" lines that help with graphing) indicate where the function isn't defined. They're lines drawn in to help when sketching the graph of a curve. The vertical asymptotes for the cosecant occur when the sine function is equal to 0. The cosecant is the reciprocal of the sine, putting the sine in the denominator of the equivalence. You can't have a 0 in the denominator. The sine is 0 for all multiples of π, so that's where the vertical asymptotes of the cosecant are: $x = k\pi$ (where k is any integer).

The vertical asymptotes for the secant occur where the cosine is equal to 0. Their equations are: $x = \frac{k\pi}{2}$, where k is an odd integer.

Q. Sketch a graph showing the vertical asymptotes for $y = \sec x$ from $x = -2\pi$ to $x = 2\pi$.

A. Your answer should look like what you see in the figure.

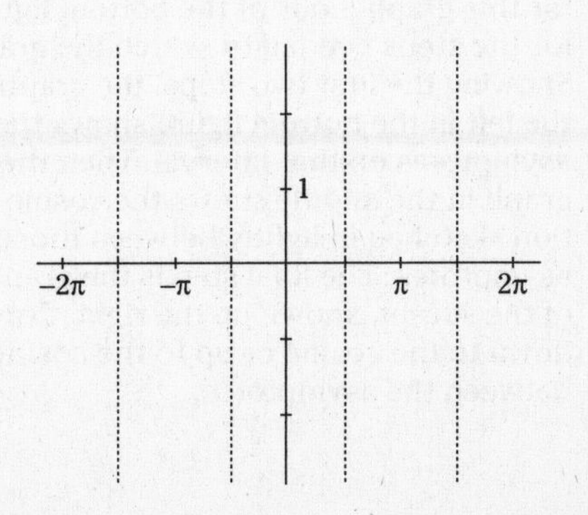

1. Give the equations of the asymptotes for $y = \csc x$ from -3π to 3π.

Solve It

2. Give the equations of the asymptotes for $y = \sec x$ from -3π to 3π.

Solve It

Graphing Cosecant and Secant

Graphing the cosecant and secant functions is a two-step process. First, sketch in the asymptotes for the function, and then lightly sketch in the reciprocal function for whichever you're graphing. The extreme values of the reciprocal functions will be the tops and bottoms of the curves you draw in for secant and cosecant.

Q. Sketch the graph of $y = \sec x$ between $x = 2\pi$ and $x = 4\pi$.

A. The top figure shows the final answer for this graph. Look at the bottom figure for the steps needed to sketch the graph. Showing the first two steps, the graph on the left in the bottom figure shows the asymptotes on that interval. Then the graph in the middle shows the cosine function sketched in lightly between those asymptotes. The final step is the graph of the secant, shown on the right, drawn down to the cosine or up to the cosine between the asymptotes.

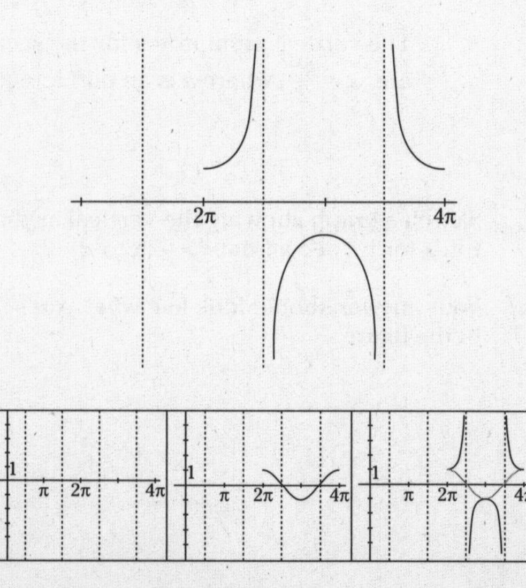

3. Sketch the graph of $y = \sec x$ from $x = -2\pi$ to $x = 2\pi$.

Solve It

4. Sketch the graph of $y = \csc x$ from $x = -2\pi$ to $x = 2\pi$.

Solve It

Making Changes to the Graphs of Cosecant and Secant

The graphs of the cosecant and secant functions look very much alike. The things that make them different from one another are where the asymptotes and extreme points are. Just as there are identities and shifts that can change from sine to cosine or vice versa, the same type of changing can be done with the cosecant and secant.

Q. Graph $y = \sec x$ and $y = \csc\left(x + \frac{\pi}{2}\right)$. What is true of the graphs?

A. These are the same graph, as shown in the figure. Moving the graph of $y = \csc x$ to the left by $\frac{\pi}{2}$ units makes it coincide with $y = \sec x$.

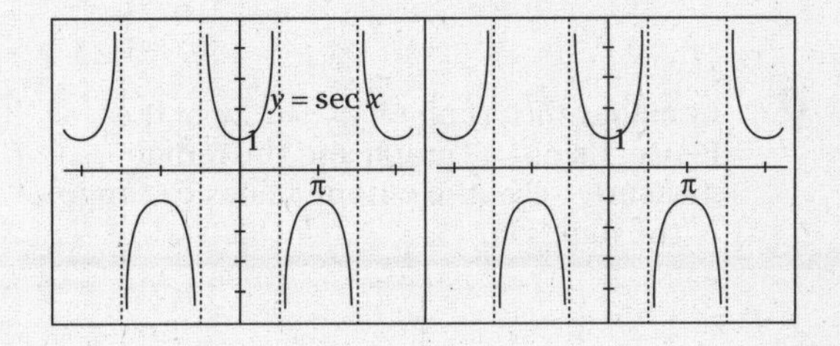

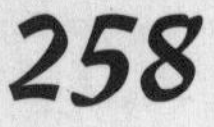

5. Sketch the graph of $y = \sec(x + 2\pi)$. What do you observe?

Solve It

6. Sketch the graph of $y = \csc\left(x + \frac{3\pi}{2}\right)$. What do you observe?

Solve It

Analyzing the Graphs of the Inverse Trig Functions

The inverse trig functions have domains that consist of all the output values of their corresponding inverses and ranges that consist of angle measures that come from two of the four quadrants. For instance, $y = \text{Sin}^{-1}x$ has input values between –1 and 1, including –1 and 1, because those are the values that $y = \sin x$ produces. The output values of this inverse function are angles from Quadrant I and Quadrant IV. There's only one output for each input — as is befitting a function. The graphs of the inverse trig functions are pretty stunted. They don't go on and on as their counterparts — their inverses — do.

Q. Consider the graph of $y = \text{Sin}^{-1}x$ in the figure. Using the graph and the radian measures, give the output values or range of the function.

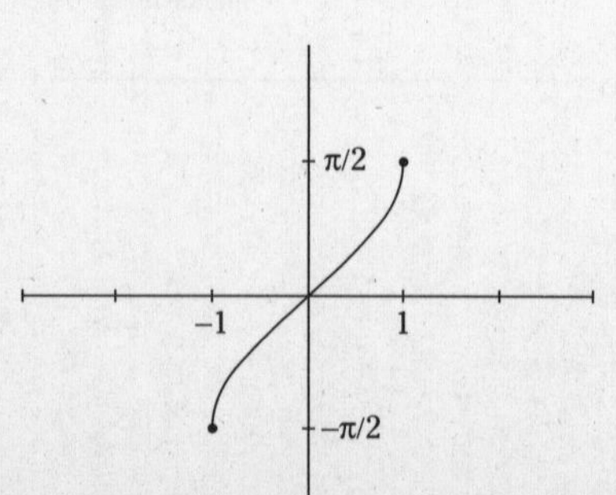

A. The range, or y values, goes from a low of $-\frac{\pi}{2}$ to a high of $\frac{\pi}{2}$. The $-\frac{\pi}{2}$ is the same as an angle of $\frac{3\pi}{2}$. Refer to Chapter 3 for information on negative angles and their positive counterparts.

7. Consider the graph of $y = \text{Cos}^{-1} x$ in the figure. Using the graph and the given measures, give the input values (domain) and output values (range) of the function.

Solve It

8. Consider the graph of $y = \text{Tan}^{-1} x$ in the figure. Using the graph and the given measures, give the input values (domain) and output values (range) of the function.

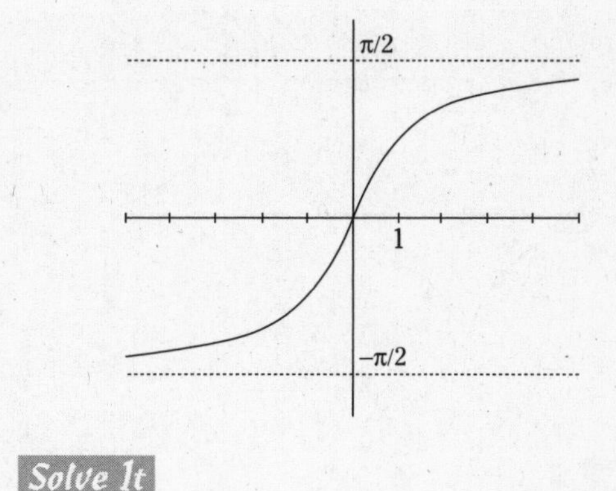

Solve It

9. Consider the graph of $y = \text{Cot}^{-1} x$ in the figure. Using the graph and the given measures, give the input values (domain) and output values (range) of the function.

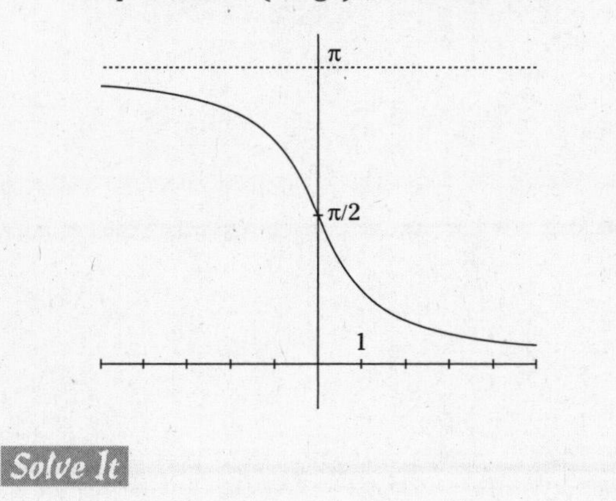

Solve It

10. Consider the graph of $y = \text{Sec}^{-1} x$ in the figure. Using the graph and the given measures, give the input values (domain) and output values (range) of the function.

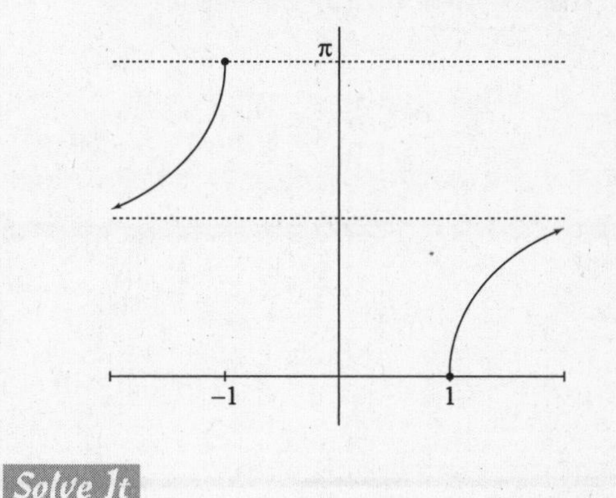

Solve It

11. Consider the graph of $y = \text{Csc}^{-1} x$ in the figure. Using the graph and the given measures, give the input values (domain) and output values (range) of the function.

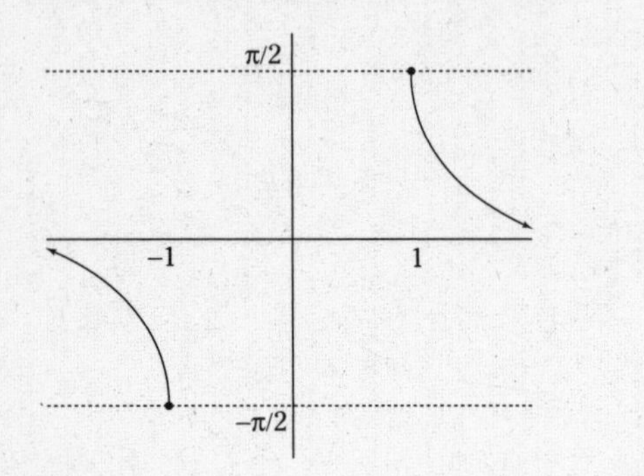

Solve It

12. Look at all the figures in this section, including the figure shown in the example question. Which have the same ranges (*y*-values)?

Solve It

Answers to Problems on Cosecant, Secant, and Inverse Trig Functions

The following are solutions to the practice problems presented earlier in this chapter.

1 Give the equations of the asymptotes for $y = \csc x$ from -3π to 3π. $\mathbf{x = -3\pi, x = -2\pi, x = -\pi, x = 0, x = \pi, x = 2\pi, x = 3\pi}$.

The asymptotes occur where the sine is equal to 0 — at multiples of π.

2 Give the equations of the asymptotes for $y = \sec x$ from -3π to 3π.

$\mathbf{x = -\frac{5\pi}{2}, x = -\frac{3\pi}{2}, x = -\frac{\pi}{2}, x = \frac{\pi}{2}, x = \frac{3\pi}{2}, x = \frac{5\pi}{2}}$.

The asymptotes occur where the cosine is equal to 0 — at odd multiples of $\frac{\pi}{2}$.

3 Sketch the graph of $y = \sec x$ from $x = -2\pi$ to $x = 2\pi$. **See the following figure.**

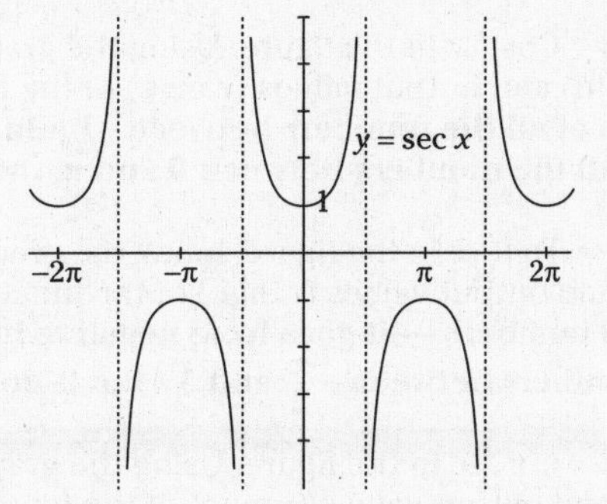

4 Sketch the graph of $y = \csc x$ from $x = -2\pi$ to $x = 2\pi$. **See the following figure.**

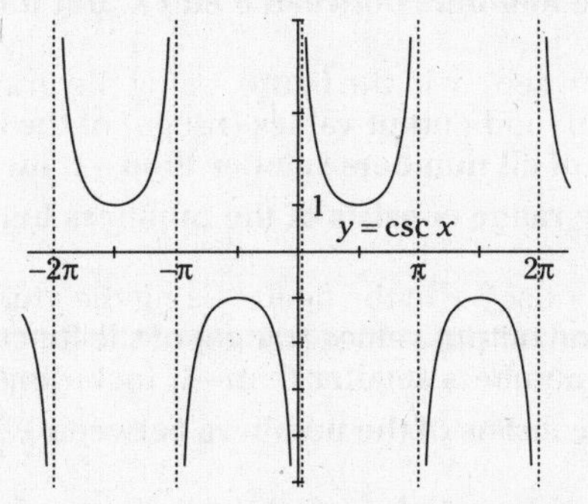

5 Sketch the graph of $y = \sec (x + 2\pi)$. What do you observe? **See the following figure.**

This is the same as the graph of $y = \sec x$. The period of the secant function is 2π, so this graph is moved on top of the original, because it moves 2π units to the left.

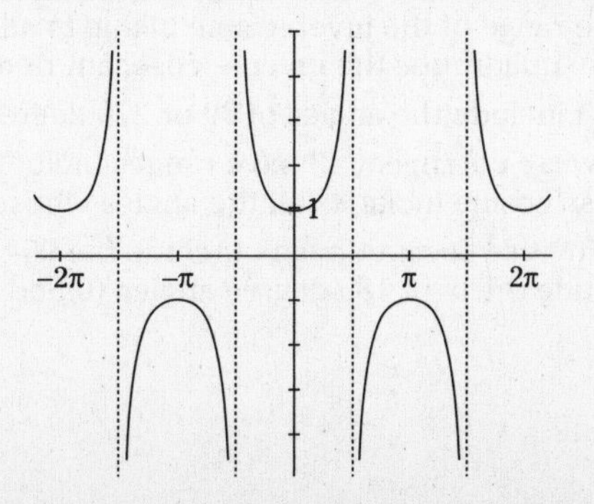

6 Sketch the graph of $y = \csc\left(x + \frac{3\pi}{2}\right)$. What do you observe? **See the following figure.**

This graph has the same asymptotes as the graph of $y = \sec x$, but it isn't the graph of the secant. This graph is actually the secant moved π units to the right, so it is equivalent to $y = \sec(x - \pi)$.

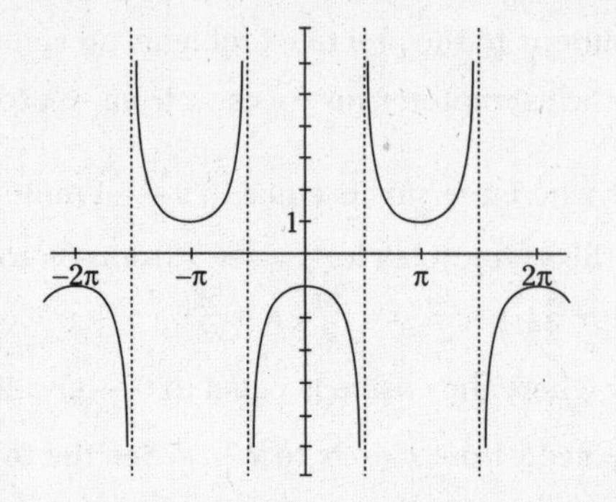

7 Consider the graph of $y = \text{Cos}^{-1}x$ in the figure. Using the graph and the given measures, give the input values (domain) and output values (range) of the function. **The domain of the inverse cosine consists of all the numbers between –1 and 1, including those two values. The range consists of all the numbers between 0 and π, including those two values.**

8 Consider the graph of $y = \text{Tan}^{-1}x$ in the figure. Using the graph and the given measures, give the input values (domain) and output values (range) of the function. **The domain of the inverse tangent consists of all real numbers — it goes from negative infinity to positive infinity. The range consist of the numbers between $-\frac{\pi}{2}$ and $\frac{\pi}{2}$, but it doesn't include those two values.**

9 Consider the graph of $y = \text{Cot}^{-1}x$ in the figure. Using the graph and the given measures, give the input values (domain) and output values (range) of the function. **The domain of the inverse cotangent consists of all real numbers — it goes from negative infinity to positive infinity. The range consist of the numbers between 0 and π, but it doesn't include those two values.**

10 Consider the graph of $y = \text{Sec}^{-1}x$ in the figure. Using the graph and the given measures, give the input values (domain) and output values (range) of the function. **The domain of the inverse secant consists of all numbers smaller than –1, including –1, and all numbers larger than 1, including 1. The range consists of the numbers between 0 and π, excluding $\frac{\pi}{2}$.**

11 Consider the graph of $y = \text{Csc}^{-1}x$ in the figure. Using the graph and the given measures, give the input values (domain) and output values (range) of the function. **The domain of the inverse cosecant consists of all numbers smaller than –1, including –1, and all numbers larger than 1, including 1. The range consists of the numbers between $-\frac{\pi}{2}$ and $\frac{\pi}{2}$, excluding 0.**

12 Look at all the figures in this section, including the figure shown in the example question. Which have the same ranges (*y*-values)? **None of the inverse trig functions has exactly the same range as any of the others, although they share many of the same angles.**

The inverse sine, inverse cosecant, and inverse tangent all have ranges including angles in Quadrants I and IV. The range of the inverse sine includes all the angles whose terminal sides are on the axes in those quadrants; the inverse cosecant doesn't include an angle of 0, and the inverse tangent doesn't include the angles of 90 or 270 degrees ($\frac{\pi}{2}$ or $-\frac{\pi}{2}$). The inverse cosine, inverse secant, and inverse cotangent all have ranges including angles in Quadrants I and II. The range of the inverse cosine includes all the angles whose terminal sides are on the axes in those quadrants. The inverse secant doesn't include the 90-degree angle ($\frac{\pi}{2}$), and the inverse cotangent doesn't include the 0- or 180-degree angles (0 and π).

Chapter 18

Transforming Graphs of Trig Functions

In This Chapter

- Sliding graphs all over the place
- Reflecting over axes
- Combining transformations

All sorts of transformations can be performed on the graphs of functions. In Chapter 15, you can read about how to make the graphs of sine and cosine slide all over the place or get taller. In Chapters 16 and 17, you can see how a *phase change* (changing the starting place) can happen when numbers are added to the angle variable. In this chapter, all the possible transformations that are performed on all the trig functions are put into one, neat package and related to the functions in general. Also, you'll see what happens when a *foreign* function is introduced — when a polynomial is added to a trig function.

Sliding the Graphs Left or Right

The function transformation called a *slide* or *translation* moves a function around the coordinate axes without changing the shape. It can change where the function has intercepts or where it starts its cycle (in the case of these trig functions). The transformations that slide the functions *left* have the form $f(x + k)$ and those that slide the functions to the *right* have the form $f(x - k)$, where k is some positive value and f is whatever function is being transformed.

Q. Sketch the graph of $y = \cot\left(x - \frac{\pi}{2}\right)$.

A. This transformation slides the cotangent function to the right by $\frac{\pi}{2}$ units. This means that the asymptotes move that amount, also. Instead of the cotangent's usual asymptotes at $x = -\pi$, $x = 0$, $x = \pi$, $x = 2\pi$, and so on, they're all moved to the right and are now $x = -\frac{\pi}{2}$, $x = \frac{\pi}{2}$, $x = \frac{3\pi}{2}$, $x = \frac{5\pi}{2}$. The figure is a sketch of the graph.

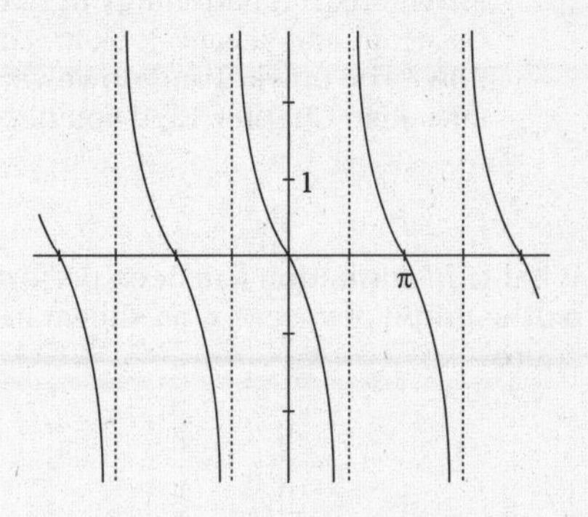

1. Sketch the graph of $y = \tan\left(x - \frac{\pi}{2}\right)$.

Solve It

2. Sketch the graph of $y = \cot\left(x + \frac{\pi}{3}\right)$.

Solve It

Sliding the Graphs Up or Down

The previous section shows you the *translations* or *slides* that move a function to the left or right. The other type of slide moves the graph up or down. The transformations that slide the functions *up* have the form $f(x) + k$, and those that slide the functions *down* have the form $f(x) - k$, where k is some positive value. The best way to graph these — or to determine that the transformation has occurred — is to look for a special or identifiable place on the graph. In the case of the tangent and cotangent, look for where that flattening-out place is. Has it moved up or down? And, in the case of the cosecant and secant, look for the tops and bottoms of the curves. Where are they now? The upward and downward movements of the sine and cosine functions are discussed in Chapter 15, if you need to refer to them.

Q. What transformation has been performed on the graph of $y = \csc x$, as shown in the figure?

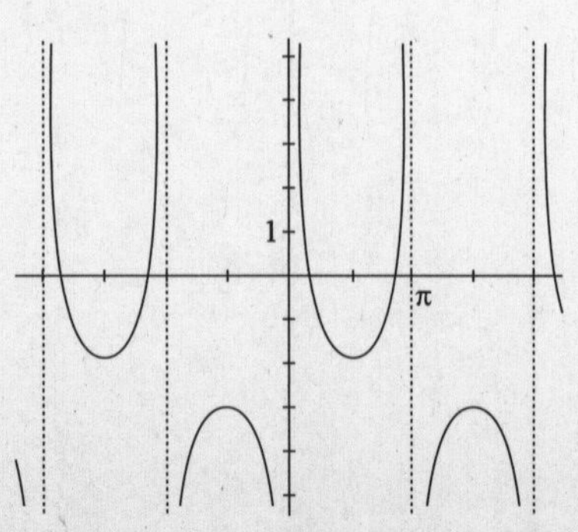

A. The graph has dropped by 2 units. Instead of having the bottoms of the upper curves at $y = 1$ and the tops of the lower curves at $y = -1$, they're at -1 and -3, respectively. The transformation is represented by the equation $y = \csc x - 2$.

3. Determine the transformation that has been performed on the graph of $y = \sec x$ (see the figure).

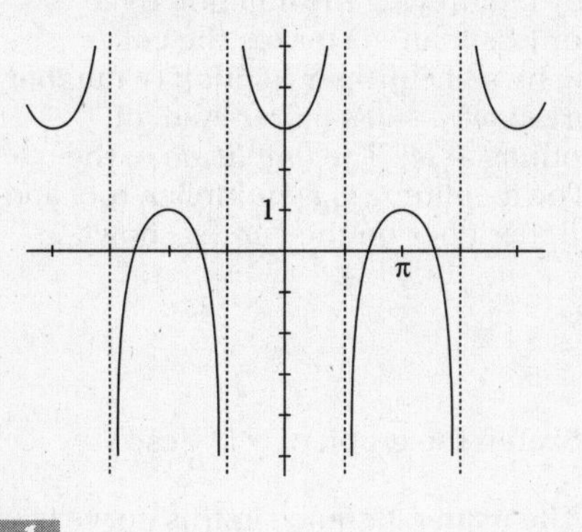

Solve It

4. Determine the transformation that has been performed on the graph of $y = \tan x$ (see the figure).

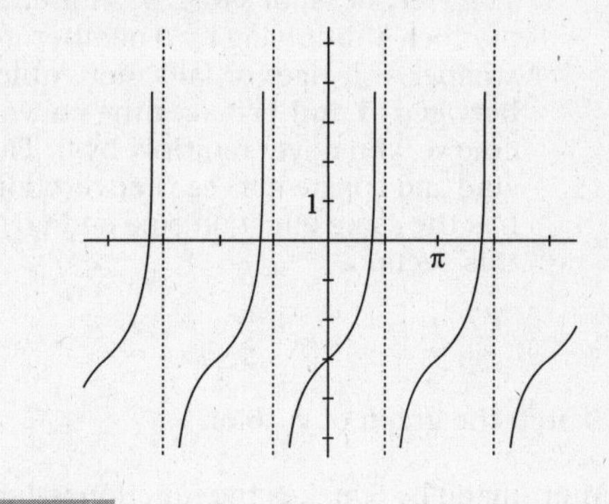

Solve It

5. Determine the transformation that has been performed on the graph of $y = \cot x$ (see the figure).

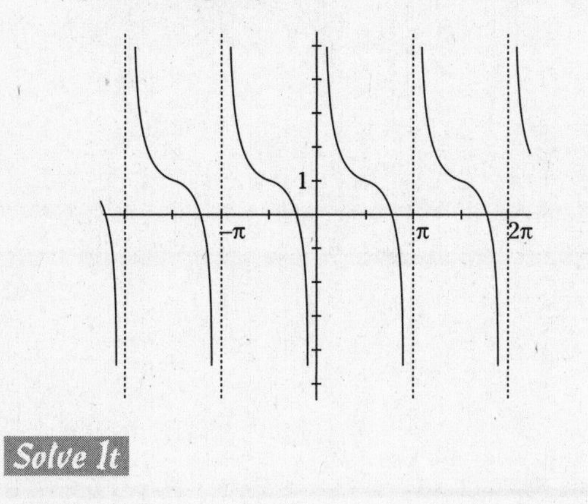

Solve It

6. Determine the transformation that has been performed on the graph of $y = \csc x$ (see the figure).

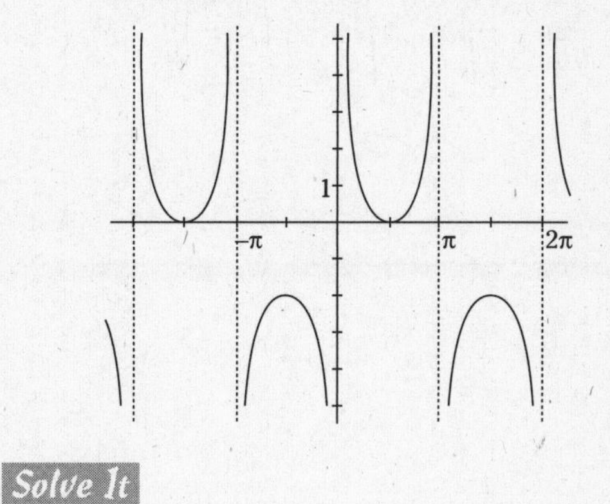

Solve It

Changing the Steepness

The steepness, or *slope,* of a function is changed by multiplying the function by a number. Multiplying by a number greater than 1 (or less than –1) makes the curve steeper — it rises or falls more quickly. Multiplying by some proper fraction (a number between –1 and 1) makes the curve rise or fall more slowly — it's flatter. And, of course, you never multiply by 0. That gives you nothing at all. The *amplitude* of the sine and cosine curves is covered in Chapter 15. The amplitude is a multiplier, too, and has the same effect on sine and cosine as the multiplier does on the curves shown in this section.

Q. Sketch the graph of $y = 6\tan x$.

A. Multiplying by 6 makes the function values six times as great. Instead of having a tangent equaling 1 unit at 45 degrees, it's now 6 units. As the numbers you multiply by get larger, the curve gets steeper and looks less like a tangent (see the figure).

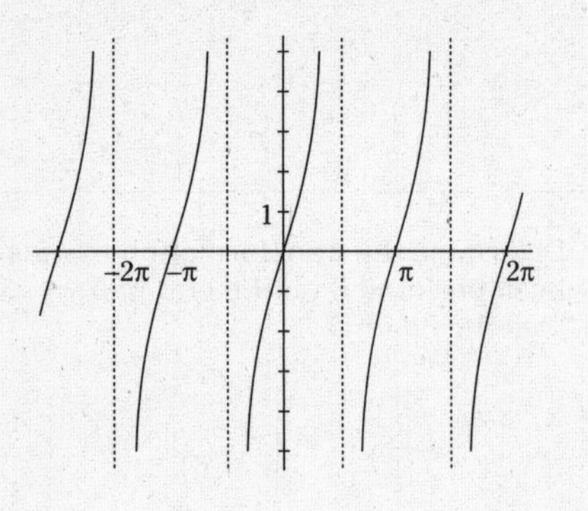

Q. Sketch the graph of $y = \frac{1}{6}\csc x$.

A. The main difference in this curve is with the values close to where the lowest part of the upper curve and highest part of the lower curve are. They flatten out and are closer to the *x*-axis because of the multiplier (see the figure).

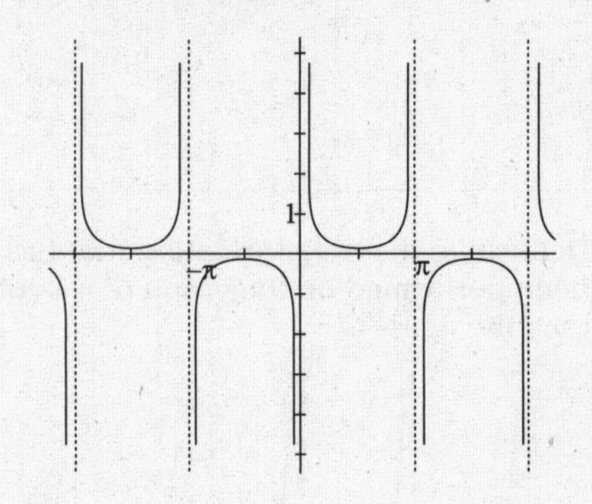

7. Sketch the graph of $y = 4\cot x$.

Solve It

8. Sketch the graph of $y = \frac{1}{3}\sec x$.

Solve It

Reflecting on the Situation — Horizontally

Another transformation that can be performed on functions is reflecting them over the *x*-axis or some other horizontal line. When you see a negative sign in front of the function rule, as in $y = -f(x)$, every *y* value that was positive becomes negative and vice versa. When you see a negative sign in a rule such as $y = -f(x)+k$, the reflection is over the line $y = k$. Graphically, the graph looks like a mirror image above and below the line it's reflecting over.

Q. What is the function rule for the graph in the figure?

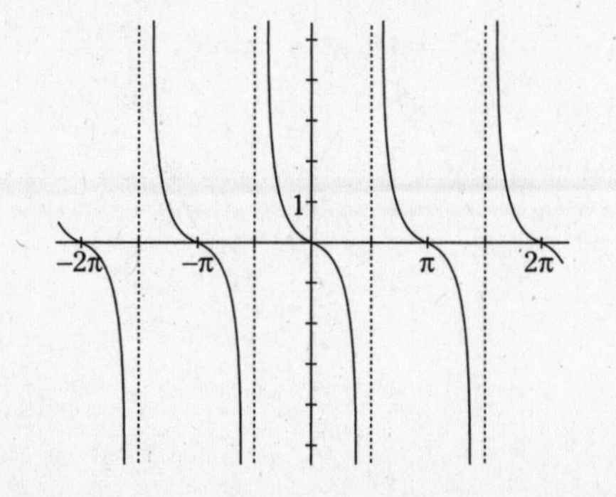

A. This is a graph of $y = -\tan x$. The graph of $y = \tan x$ has been reflected over the *x*-axis.

Q. What is the function rule for the graph in the figure?

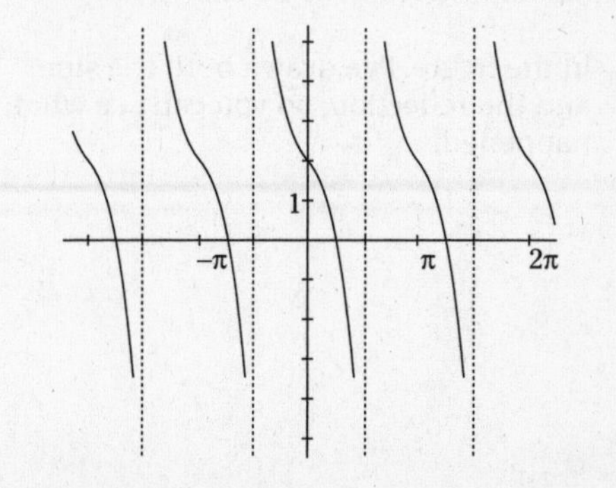

A. This is the tangent function reflected again. But this time, it's been raised by 2 units also. So it's been reflected over the line $y = 2$. The function equation is $y = -\tan x + 2$.

9. Determine the function equation for the graph in the figure using a reflection.

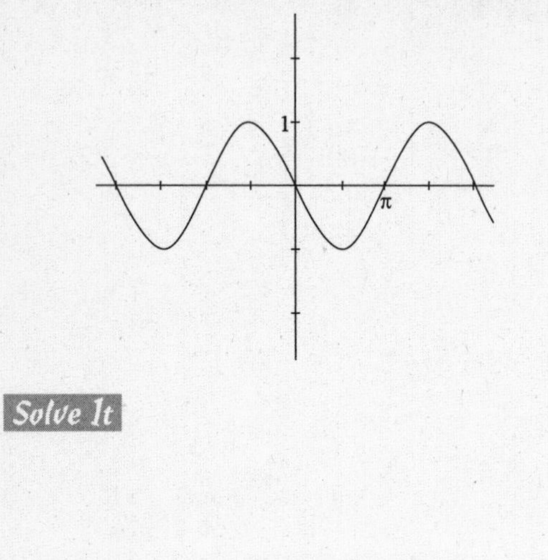

Solve It

10. Determine the function equation for the graph in the figure.

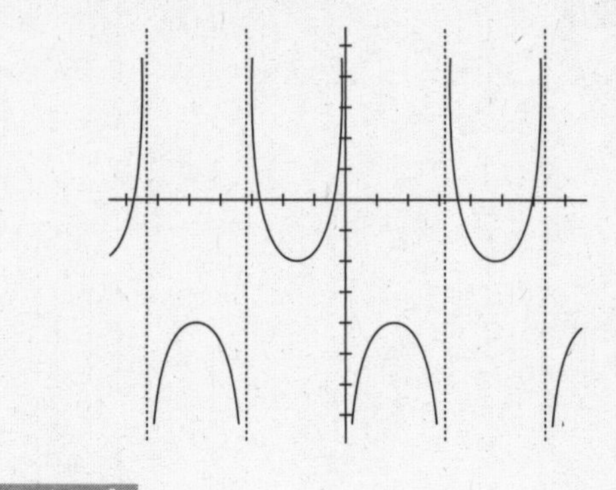

Solve It

Reflecting on Your Position — Vertically

There are two types of reflections in the world of transformations. Reflections over a horizontal line are covered in the preceding section. The others, covered here, are reflections over a vertical line — most often the y-axis. The function equation for these transformations has the variable for the angle negated. You'll see $f(-x)$. Everything switches sides from left to right and right to left. The functions that are already symmetric with respect to the y-axis (they're a mirror image on either side) can be reflected, but you'll never know it; they reflect back on themselves.

Q. Sketch the graph of $y = \sin(-x)$.

A. In the figure, I've drawn both the sine and the reflection, so you can see what happened.

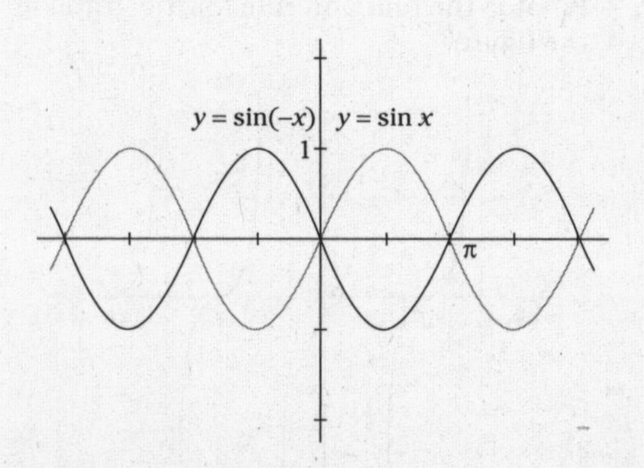

11. Sketch the graph of $y = \tan(-x)$.

Solve It

12. Sketch the graph of $y = \csc(-x)$.

Solve It

Putting It All Together

Transformations can be performed on functions that raise, lower, shift left, shift right, steepen, flatten, and flip it. All of these are described in the sections earlier in this chapter. For the most part, in those earlier sections, only one transformation has been applied each time. In this section, I show you how two or more transformations can affect the function.

Q. Sketch the graph of $y = \frac{1}{3}\cot\left(x - \frac{\pi}{2}\right) + 1$.

A. This graph is really loaded. It's flatter, because of the fraction multiplier. And it's shifted to the right by $\frac{\pi}{2}$ units. The figure shows you the end result.

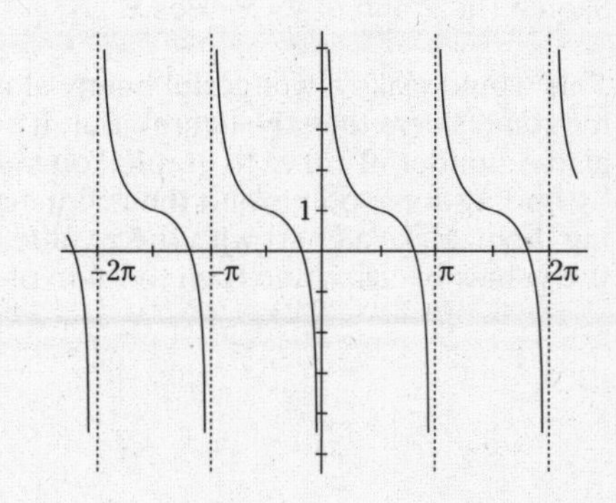

13. Sketch the graph of $y = 3\sec x + 2$.

Solve It

14. Sketch the graph of $y = \frac{1}{4}\tan\left(x + \frac{\pi}{2}\right) + 3$.

Solve It

Combining Trig Functions with Polynomials

Some of the prettiest (beauty is in the eye of the beholder) graphs are achieved when you combine two different trig functions or a trig function and a polynomial together. Graphing calculators come in handy here, but you can also do a good job of combining these functions by just doing the math yourself and plotting the points.

Q. Sketch the graph of $y = x + \cos x$.

A. This would make a wonderful bunny slope for some skiers (see the figure). But, it's also a wonderful curve to graph. You start by finding some points and then connecting them. Make a chart with the x-values, the values of $\cos x$, and then the sum of those two (see the chart). The radian measures are changed to decimals for the addition. You don't get a feel for the smoothness without trying some points in between the points in the chart, but the rises and plateaus are completely predictable with the cosine function having those values between –1 and 1.

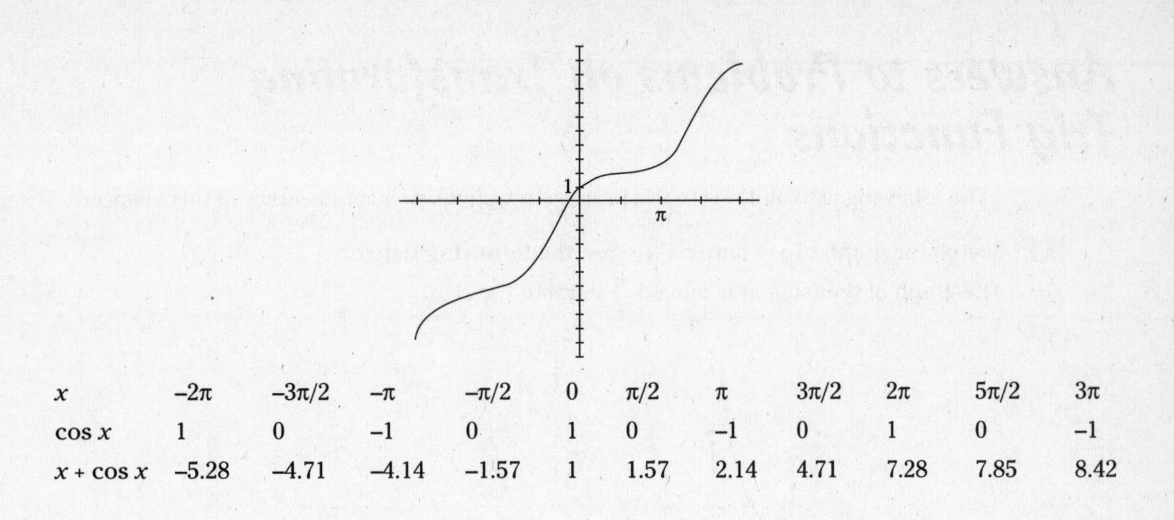

x	-2π	$-3\pi/2$	$-\pi$	$-\pi/2$	0	$\pi/2$	π	$3\pi/2$	2π	$5\pi/2$	3π
$\cos x$	1	0	–1	0	1	0	–1	0	1	0	–1
$x + \cos x$	–5.28	–4.71	–4.14	–1.57	1	1.57	2.14	4.71	7.28	7.85	8.42

15. Sketch the graph of $y = x + 2 \sin x$.

Solve It

16. Sketch the graph of $y = 3\cos x - 2x$.

Solve It

Answers to Problems on Transforming Trig Functions

The following are solutions to the practice problems presented earlier in this chapter.

1 Sketch the graph of $y = \tan\left(x - \frac{\pi}{2}\right)$. **See the following figure.**

The graph of the tangent is moved $\frac{\pi}{2}$ units to the right.

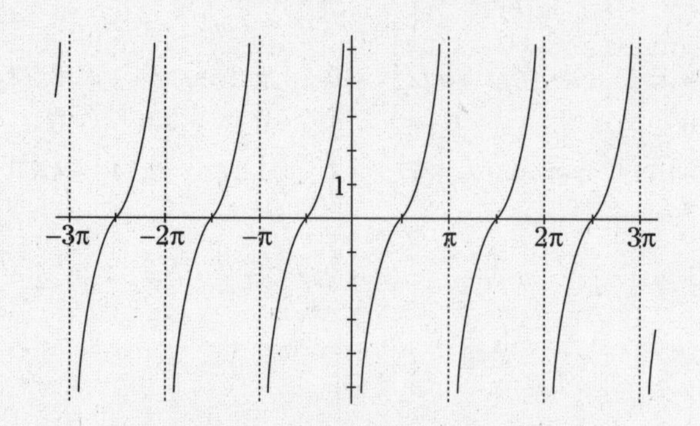

2 Sketch the graph of $y = \cot\left(x + \frac{\pi}{3}\right)$. **See the following figure.**

The graph of the cotangent is moved $\frac{\pi}{3}$ units to the left.

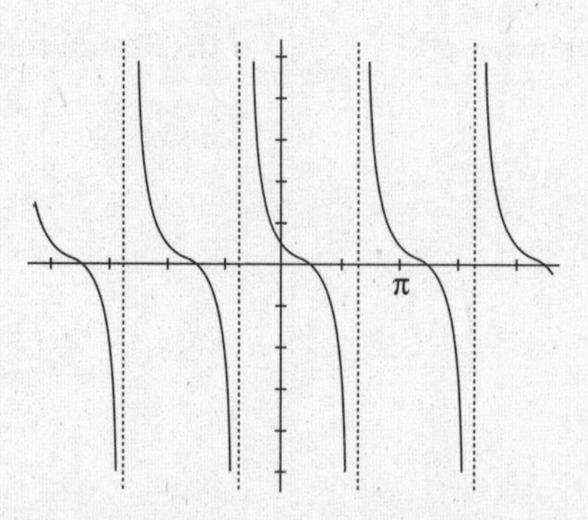

3 Determine the transformation that has been performed on the graph of $y = \sec x$ (see the figure). **This graph is 2 units higher than $y = \sec x$.**

Its function equation is $y = \sec x + 2$.

4 Determine the transformation that has been performed on the graph of $y = \tan x$ (see the figure). **This graph is 3 units lower than $y = \tan x$.**

Its function equation is $y = \tan x - 3$.

5 Determine the transformation that has been performed on the graph of $y = \cot x$ (see the figure). **This graph is 1 unit higher than $y = \cot x$.**

Look for where there's the bend in the curve, and see where the new bend is, relative to where it usually is. Its function equation is $y = \cot x + 1$.

6 Determine the transformation that has been performed on the graph of $y = \csc x$ (see the figure). **This graph is 1 unit lower than $y = \csc x$.**

Its function equation is $y = \csc x - 1$.

7 Sketch the graph of $y = 4\cot x$. **See the following figure.**

This is much steeper than $y = \cot x$. The multiplier of 4 is what makes it steeper. Notice, though, that the intercepts are the same. That's because 0 times 4 is still 0. The asymptotes stay the same, too.

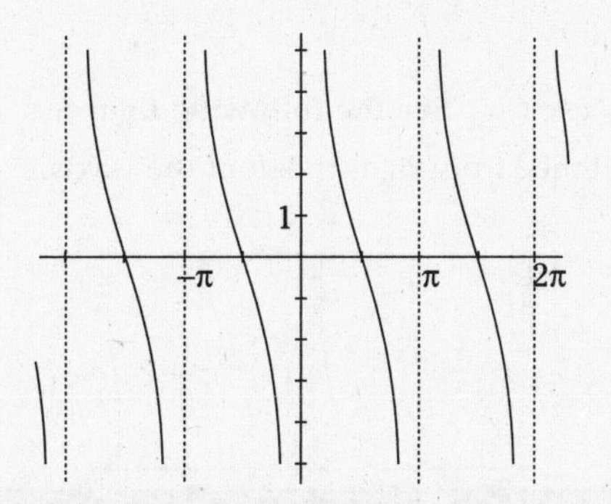

8 Sketch the graph of $y = \frac{1}{3}\sec x$. **See the following figure.**

The curve becomes flatter, and the upper and lower portions of the curve come closer to the *x*-axis.

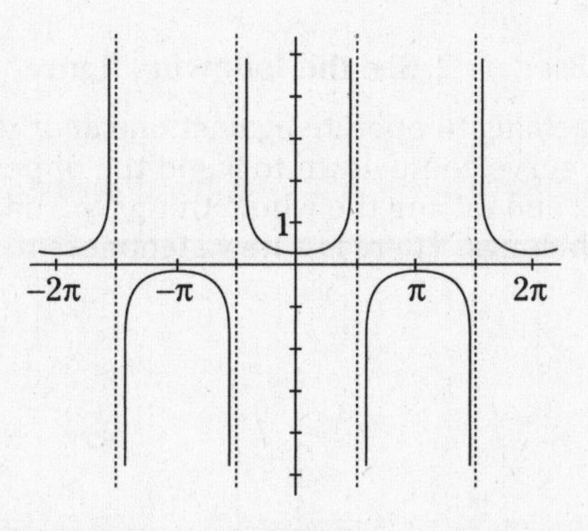

9 Determine the function equation for the graph in the figure using a reflection. **This is the sine curve reflected over the *x*-axis. The equation is $y = -\sin x$.**

10 Determine the function equation for the graph in the figure. **This is the cosecant curve that's been dropped by 3 units and then reflected over the line $y = -3$. The equation is $y = -\csc x - 3$.**

11 Sketch the graph of $y = \tan(-x)$. **See the following figure.**

All the points have switched from right to left of the y-axis.

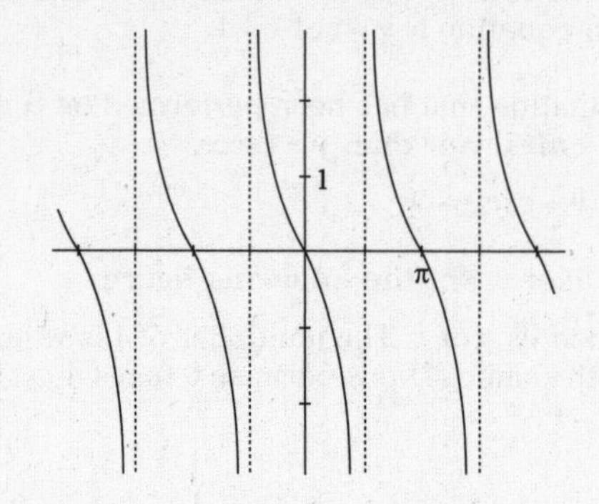

12 Sketch the graph of $y = \csc(-x)$. **See the following figure.**

All the points have switched from right to left of the y-axis.

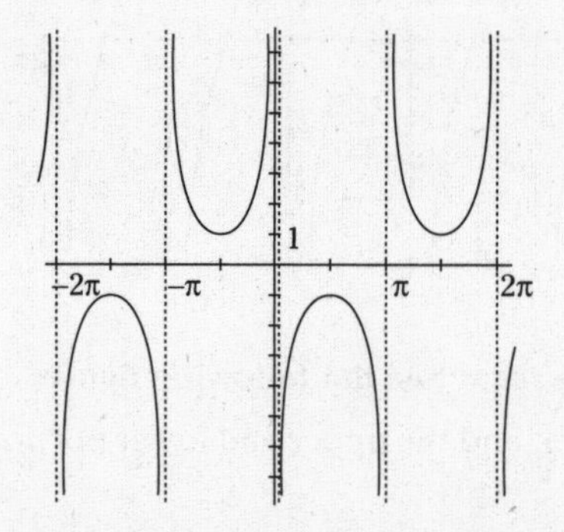

13 Sketch the graph of $y = 3\sec x + 2$. **See the following figure.**

There are two factors seeming to operate against one another here. Multiplying by 3 makes the lower part of the upper curve come down to 3 and the upper part of the lower curve come up to –3. But then, adding 2 and raising the whole thing by 2 units makes the lower part seem to come back up where it belongs. There is a new steepness to the curve. The multiplier of 3 does make it steeper.

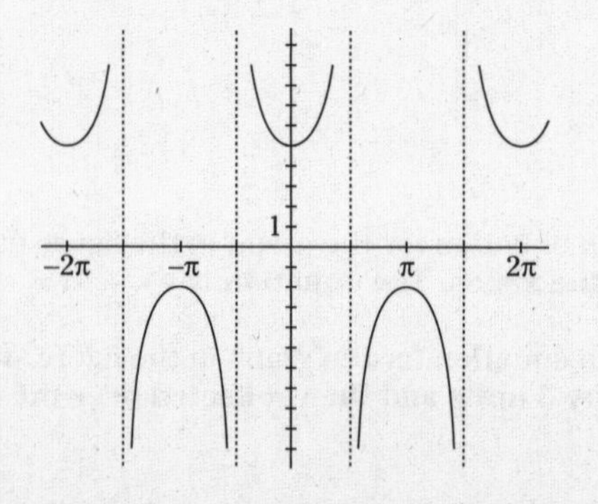

 Sketch the graph of $y = \frac{1}{4}\tan\left(x + \frac{\pi}{2}\right) + 3$. **See the following figure.**

There's a lot going on here. The curve is raised 3 units; it's shifted to the left by $\frac{\pi}{2}$ units; and then it's flattened out by the fractional multiplier. Notice that the asymptotes have shifted with the curve.

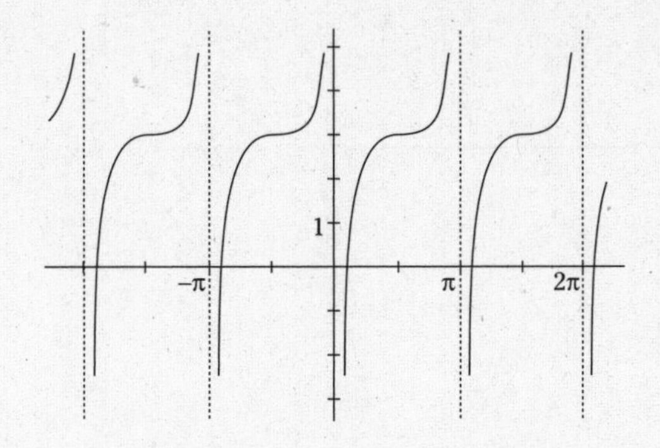

15 Sketch the graph of $y = x + 2\sin x$. **See the following figure.**

By adding function values, you can get a pretty good idea of how this curve is shaping out.

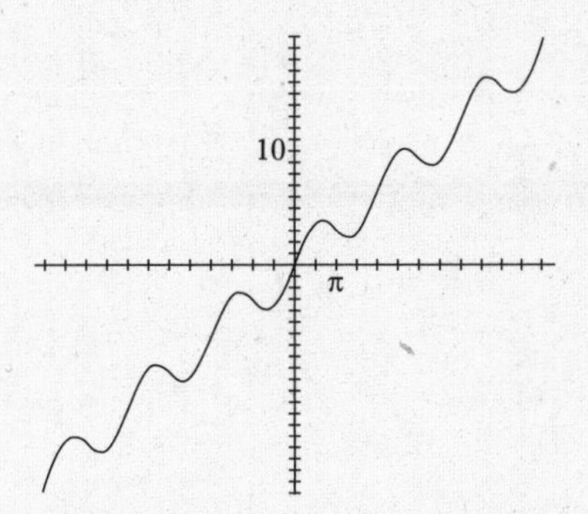

16 Sketch the graph of $y = 3\cos x - 2x$. **See the following figure.**

Choose enough function values and find their difference to see what this looks like.

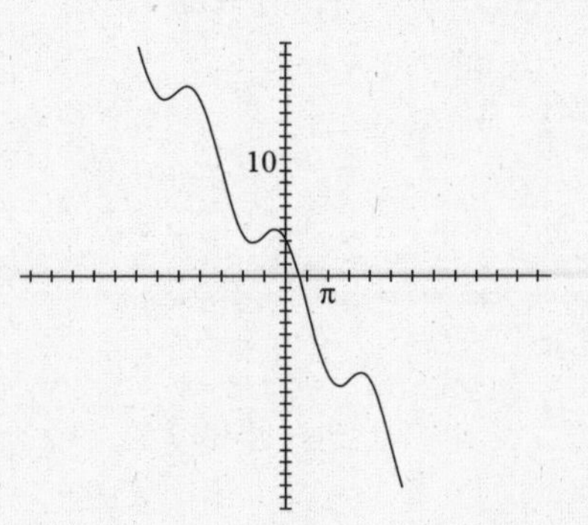

Part V
The Part of Tens

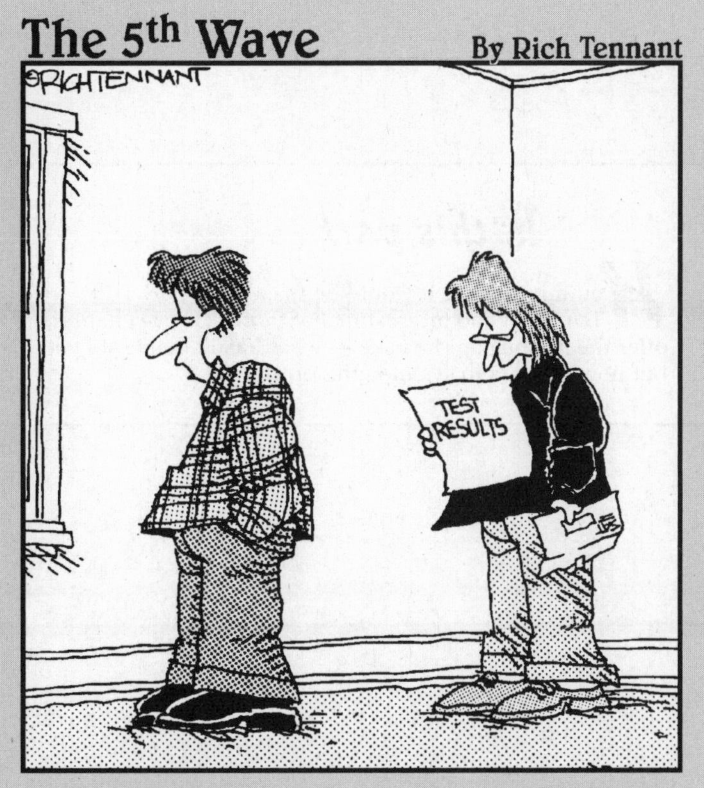

"When my parents see the grade I got in trigonometry, they're going to have a Pythagorean fit."

In this part . . .

How appropriate it would be to call this part TGIF: Ten Great and Interesting Footnotes. These chapters offer the frosting on the cake — what hasn't been said yet but now appears to sweeten the product.

Chapter 19

Ten Identities with a Negative Attitude

In This Chapter

- Soothing negative angle identities
- Complementing and supplementing angles
- Factoring in some positive approaches
- Changing the demeanor of angles

There are oh-so-many identities. In Part III of this book, you find the most commonly used identities and see how they can be used to simplify trig expressions and solve trig equations. Here I list ten identities, three of which you may recognize, and the others of which crop up in graphing and simplifying situations.

Negative Angle Identities

There are three negative angle identities that tell you what the function of a negative angle is with respect to the corresponding positive angle. For example, if you want to find the sine of an angle of –45 degrees, you can find it by just negating the sine of a 45 degree angle that's positive. Here are the identities:

$$\sin(-x) = -\sin x$$

$$\cos(-x) = \cos x$$

$$\tan(-x) = -\tan x$$

Complementing and Supplementing Identities

This next grouping of identities relates the *complement* of an angle (complementary angles add up to 90 degrees) and relates it to a *supplement* of an angle (supplements add up to 180 degrees). Complements are very important in right-triangle trig, because the two acute angles in a right triangle are always complementary. The supplements are great when dealing with angles that lie along the same straight line.

$$\sin x = \cos(90 - x) = \sin(180 - x)$$

$$\cos x = \sin(90 - x) = -\cos(180 - x)$$

$$\tan x = \cot(90 - x) = -\tan(180 - x)$$

$$\cot x = \tan(90 - x) = -\cot(180 - x)$$

Doing Fancy Factoring with Identities

The Pythagorean identities do a pretty good job of covering all the different ways that those equations can be written and substituted in and used in solving equations. These last identities take the difference between the squares of sine and cosine functions and, instead of factoring, relate them to some sum and difference identities.

$$\sin^2 x - \sin^2 y = \sin(x+y)\sin(x-y)$$

$$\cos^2 x - \cos^2 y = -\sin(x+y)\sin(x-y)$$

$$\cos^2 x - \sin^2 y = \cos(x+y)\cos(x-y)$$

Chapter 20

Ten Formulas to Use in a Circle

In This Chapter

- Finding your way around a circle
- Sectioning off the area of a circle
- Inscribing and circumscribing a circle

The circle is one of the most familiar and useful objects. Even young children who haven't been to school know how to form a circle around a playmate or run around in circles. The different formulas and properties that I list here may make it even more interesting and useful to you. Some of these formulas are probably familiar; many are probably not.

Running Around in Circles

The *circumference,* or distance around the outside of a circle can be found if you know the *radius* (distance from the center to the circle) or the *diameter* (distance across the circle through the center). To find the circumference, $C = 2\pi r = \pi d$ where C is the circumference, π is about 3.14, r is the radius, and d is the diameter.

Adding Up the Area

The *area* of a circle can be found if you have the radius of the circle. $A = \pi r^2$ where A is the area, π is about 3.14, and r is the radius.

Defeating an Arc Rival

An arc is a piece of a circle. It's usually measured by the angle formed between two radii of the circle and is how much of the circle is cut off. In Figure 20-1, the arc AB is a part of the whole circle. The angle that's cut off is Θ.

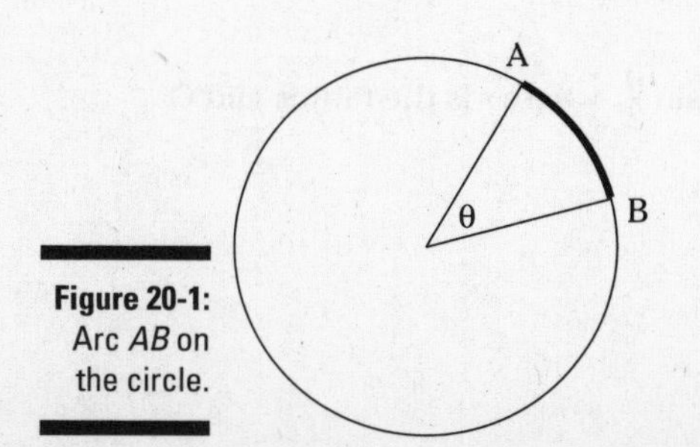

Figure 20-1: Arc *AB* on the circle.

To find the length of the arc, use the formula $s = \frac{\pi r\theta}{180}$ when Θ is in degrees and $s = r\theta$ when Θ is in radians. The letter *s* is the arc length and *r* is the radius of the circle.

Sectioning Off the Sector

The *sector* of a circle is a like a piece of a round pie. It's sandwiched between two radii of the circle. You can find the area of a sector if you know the angle of the circle that's being cut off and the radius of the circle. Figure 20-2 is a picture of a sector.

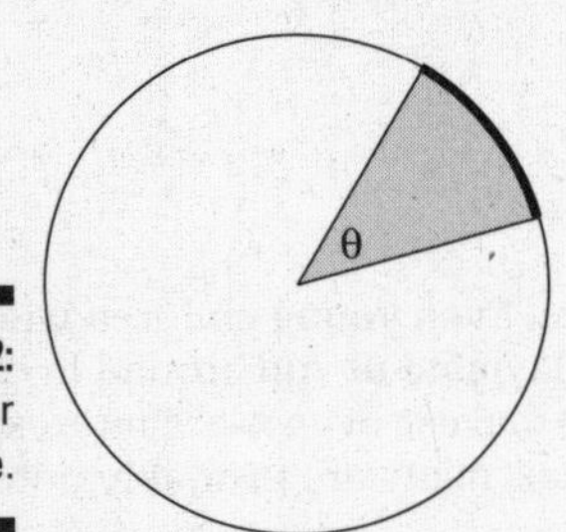

Figure 20-2: The sector of a circle.

To find the area of the sector, use $A = \frac{\pi r^2\theta}{360}$ if Θ is in degrees and $A = \frac{sr}{2} = \frac{r^2\theta}{2}$ if Θ is in radians. The *r* is for the radius, and *s* is the arc length.

Striking a Chord

A chord of a circle is a segment that's drawn from one point on the circle to another point on the circle. The longest chord is the diameter. You can quickly find this length if you know the radius of the circle and the measure of the angle that's formed from the radii touching the two ends of the chord. Look at the picture of the chord in Figure 20-3.

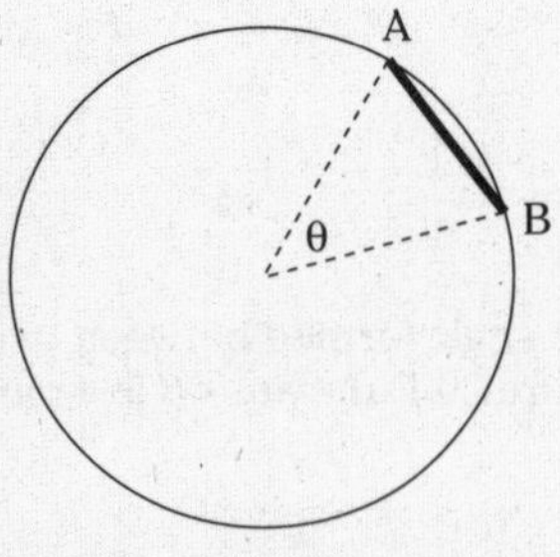

Figure 20-3: A chord *AB* drawn in a circle.

The length of the chord is found with the formula $2r\sin\frac{\theta}{2}$, where *r* is the radius and Θ is the angle between the two radii.

Ringing True

A *ring* in a circle is the area sandwiched between the circle and another circle drawn inside that has the same center. To find the area of a ring, you just need to know the radii of the two circles. Figure 20-4 shows a circle with a ring drawn inside.

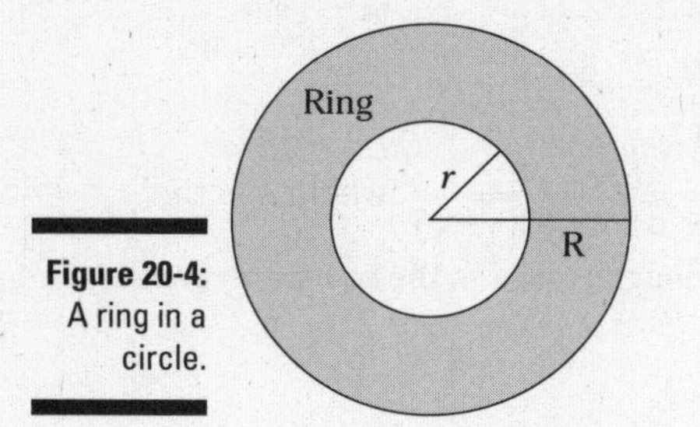

Figure 20-4: A ring in a circle.

The area of a ring is found with $A = \pi(R+r)(R-r)$, where R is the radius of the big circle, and r is the radius of the circle inside it. If you do a little algebra to this equation — multiply the two binomials together and distribute the π — you see that it's the area of the small circle subtracted from the area of the large circle. This formula is just a little more convenient.

Inscribing and Radii

A circle can be *inscribed* in a triangle. This means that the circle touches each side of the triangle it's in. You can see how that can happen in Figure 20-5. A nice thing is that you can find the radius of the circle that's inscribed inside if you know the lengths of the sides of the triangle.

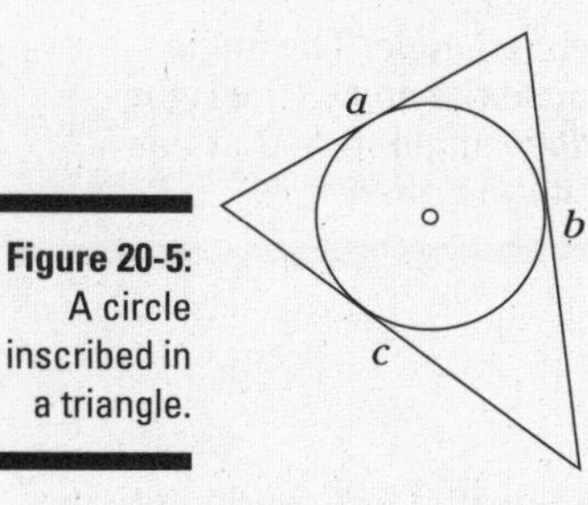

Figure 20-5: A circle inscribed in a triangle.

The radius of a circle inscribed inside a triangle that has sides measuring *a, b,* and *c* is $r = \frac{\sqrt{s(s-a)(s-b)(s-c)}}{s}$, where r is the radius and s is the *semiperimeter* (half the perimeter).

Circumscribing and Radii

A circle can also *circumscribe* a triangle. A circle can be drawn through all three vertices of any triangle (see Figure 20-6). The radius of that circle can be found if you know the lengths of the sides of the triangle.

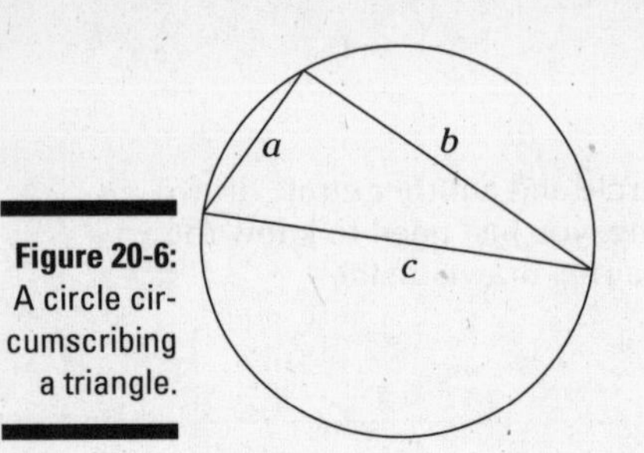

Figure 20-6: A circle circumscribing a triangle.

The formula for the radius of this circle is $R = \dfrac{abc}{4\sqrt{s(s-a)(s-b)(s-c)}}$, where R is the radius; a, b, and c are the lengths of the sides of the triangle; and s is the *semiperimeter* (half the perimeter of the triangle).

Righting a Triangle

If you draw a triangle inside a semicircle (half a circle) where one side of the triangle is the diameter, some wonderful things can happen. The triangle has a vertex at any point on the semicircle, and two endpoints of the diameter are the other two vertices. Look at Figure 20-7 for examples.

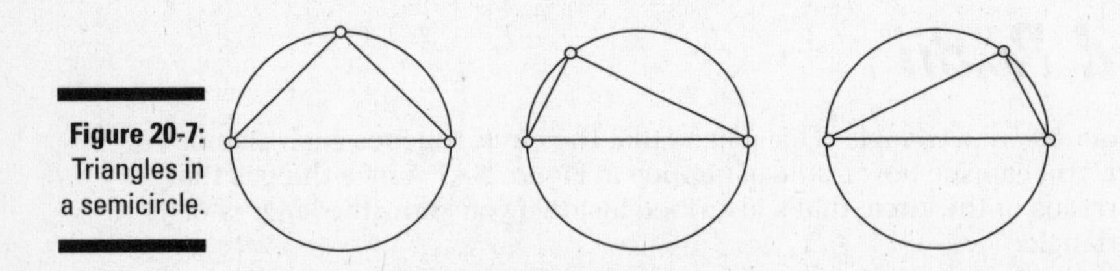
Figure 20-7: Triangles in a semicircle.

In every case, the triangle that's drawn in the semicircle is a right triangle. The angle that has its vertex on the circle is a right angle. This is true because the angle (the right angle) cuts off an arc of 180 degrees, or half the circle. An *inscribed* angle (one that has its vertex on the circle) is always half the measure of the angle it cuts off.

Inscribing a Polygon

A *polygon* is a figure made up of a bunch of segments that are connected from endpoint to endpoint. A *regular polygon* is a polygon that has all those sides or segments equal in length. When a regular polygon is inscribed inside a circle, you can find the perimeter and area of the polygon, as long as you know the radius of the circle.

To find the perimeter of the polygon, use $P = 2nr\sin\frac{\pi}{n}$, where P is the perimeter, n is the number of sides, and r is the radius of circle. To find the area of the polygon, use $A = \frac{1}{2}nr^2\sin\frac{2\pi}{n}$.

Chapter 21

Ten Ways to Relate the Sides and Angles of Any Triangle

In This Chapter

- Putting sides and angles in ratios
- Being radical with the sides
- Using trig functions in relations

The sides and angles of any triangle are related to one another in many ways. The longest side of a triangle is always opposite the largest angle, and the smallest side is opposite the smallest angle. The Law of Sines and Law of Cosines show some other relationships, but here are some you may not have seen before.

The sides of these triangles are labeled with lowercase letters, and the angles opposite them have corresponding uppercase letters. For simplicity, they'll all be for triangle *ABC*, shown in Figure 21-1.

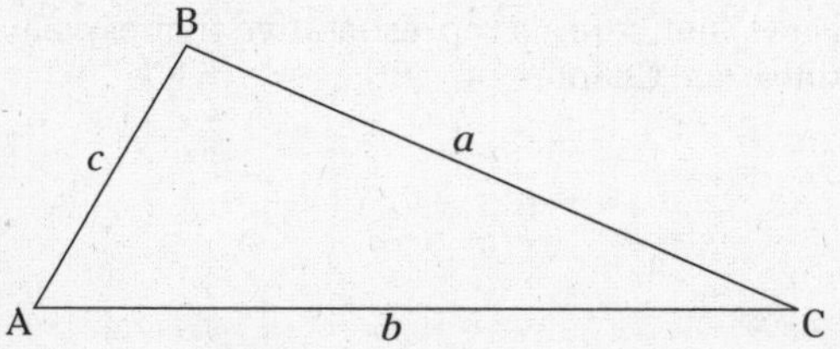

Figure 21-1: Triangle *ABC*.

Relating with the Law of Sines

This first relationship is the Law of Sines, with an added twist.

$\frac{a}{\sin A} = \frac{b}{\sin B} = \frac{c}{\sin C}$ = diameter of the circumscribed circle (the circle that can be drawn around the triangle)

Hatching a Little Heron

Heron's rule for finding the area of a triangle (see Chapter 14) uses the *semiperimeter* (half the perimeter) and the square root of a product. You can see some of Heron's rule in this relation. The s stands for the semiperimeter.

$$\sin A = \frac{2}{bc}\sqrt{s(s-a)(s-b)(s-c)}$$

Summing Sines

This next relationship relates the sums and differences of the measures of the sides to the sums and differences of the corresponding sines.

$$\frac{a+b}{a-b} = \frac{\sin A + \sin B}{\sin A - \sin B}$$

You Half It or You Don't

This relationship involving sines uses a variation on the half-angle formula.

$$\sin\frac{A}{2} = \sqrt{\frac{(s-b)(s-c)}{bc}}$$

Cozying Up with Cosines

The Law of Cosines is actually three laws, but here's a representative that explains all of them. For more on the Law of Cosines, see Chapter 14.

$$a = b^2 + c^2 - 2bc\cos A$$

Angling for an Angle

The side-side-side application of the Law of Cosines allows you to solve for the measure of an angle. To do that, solve for the cosine first, using this formula.

$$\cos A = \frac{b^2 + c^2 - a^2}{2bc}$$

Mixing It Up with Cosines

This relationship mixes up the sides and the cosines of other sides.

$$a = b\cos C + c\cos B$$

Heron Again, Gone Tomorrow

Yes, this is a revisit of Heron's rule. This time, it's the cosine that gets involved. The *s* is the semiperimeter.

$$\cos\frac{A}{2} = \sqrt{\frac{s(s-a)}{bc}}$$

Divide and Conquer with the Tangent

This relationship uses a half angle of a difference of angles. If that isn't exciting enough for you, it also multiplies the ratio of the difference and sum of sides times a cotangent half-angle. Whew!

$$\tan\frac{A-B}{2} = \frac{a-b}{a+b}\cot\frac{C}{2}$$

Heron Lies the Problem

Aren't you glad this is the last relationship, so you don't have to see any more Heron headings? This last relationship involves the tangent.

$$\tan\frac{A}{2} = \sqrt{\frac{(s-b)(s-c)}{s(s-a)}}$$

Appendix

Trig Functions Table

The values for trig functions come in handy when solving problems, sketching graphs, or doing any number of computations involving trig. The values here are all rounded to three decimal places; if you need more decimal places than that, you can use a scientific calculator. If you need values for angles larger than 90 degrees, check out Chapters 3 and 8 to see how to use coterminal and reference angles. If you need radians instead of degrees, Chapter 4 has information on changing from radians to degrees and back again.

_	*sin _*	*cos _*	*tan _*	*cot _*	*sec _*	*csc _*
0°	.000	1.000	.000	Undefined	1.000	Undefined
1°	.017	1.000	.017	57.290	1.000	57.299
2°	.035	.999	.035	28.636	1.001	28.654
3°	.052	.999	.052	19.081	1.001	19.107
4°	.070	.998	.070	14.301	1.002	14.336
5°	.087	.996	.087	11.430	1.004	11.474
6°	.105	.995	.105	9.514	1.006	9.567
7°	.122	.993	.123	8.144	1.008	8.206
8°	.139	.990	.141	7.115	1.010	7.185
9°	.156	.988	.158	6.314	1.012	6.392
10°	.174	.985	.176	5.671	1.015	5.759
11°	.191	.982	.194	5.145	1.019	5.241
12°	.208	.978	.213	4.705	1.022	4.810
13°	.225	.974	.231	4.331	1.026	4.445
14°	.242	.970	.249	4.011	1.031	4.134
15°	.259	.966	.268	3.732	1.035	3.864
16°	.276	.961	.287	3.487	1.040	3.628
17°	.292	.956	.306	3.271	1.046	3.420
18°	.309	.951	.325	3.078	1.051	3.236
19°	.326	.946	.344	2.904	1.058	3.072
20°	.342	.940	.364	2.747	1.064	2.924
21°	.358	.934	.384	2.605	1.071	2.790
22°	.375	.927	.404	2.475	1.079	2.669

(continued)

_	sin_	cos_	tan_	cot_	sec_	csc_
23°	.391	.921	.424	2.356	1.086	2.559
24°	.407	.914	.445	2.246	1.095	2.459
25°	.423	.906	.466	2.145	1.103	2.366
26°	.438	.899	.488	2.050	1.113	2.281
27°	.454	.891	.510	1.963	1.122	2.203
28°	.469	.883	.532	1.881	1.133	2.130
29°	.485	.875	.554	1.804	1.143	2.063
30°	.500	.866	.577	1.732	1.155	2.000
31°	.515	.857	.601	1.664	1.167	1.972
32°	.530	.848	.625	1.600	1.179	1.887
33°	.545	.839	.649	1.540	1.192	1.836
34°	.559	.829	.675	1.483	1.206	1.788
35°	.574	.819	.700	1.428	1.221	1.743
36°	.588	.809	.727	1.376	1.236	1.701
37°	.602	.799	.754	1.327	1.252	1.662
38°	.616	.788	.781	1.280	1.269	1.624
39°	.629	.777	.810	1.235	1.287	1.589
40°	.643	.766	.839	1.192	1.305	1.556
41°	.656	.755	.869	1.150	1.325	1.524
42°	.669	.743	.900	1.111	1.346	1.494
43°	.682	.731	.933	1.072	1.367	1.466
44°	.695	.719	.966	1.036	1.390	1.440
45°	.707	.707	1.000	1.000	1.414	1.414
46°	.719	.695	1.036	.966	1.440	1.390
47°	.731	.682	1.072	.933	1.466	1.367
48°	.743	.669	1.111	.900	1.494	1.346
49°	.755	.656	1.150	.869	1.524	1.325
50°	.766	.643	1.192	.839	1.556	1.305
51°	.777	.629	1.235	.810	1.589	1.287
52°	.788	.616	1.280	.781	1.624	1.269
53°	.799	.602	1.327	.754	1.662	1.252
54°	.809	.588	1.376	.727	1.701	1.236
55°	.819	.574	1.428	.700	1.743	1.221
56°	.829	.559	1.483	.675	1.788	1.206

_	sin _	cos _	tan _	cot _	sec _	csc _
57°	.839	.545	1.540	.649	1.836	1.192
58°	.848	.530	1.600	.625	1.887	1.179
59°	.857	.515	1.664	.601	1.972	1.167
60°	.866	.500	1.732	.577	2.000	1.155
61°	.875	.485	1.804	.554	2.063	1.143
62°	.883	.469	1.881	.532	2.130	1.133
63°	.891	.454	1.963	.510	2.203	1.122
64°	.899	.438	2.050	.488	2.281	1.113
65°	.906	.423	2.145	.466	2.366	1.103
66°	.914	.407	2.246	.445	2.459	1.095
67°	.921	.391	2.356	.424	2.559	1.086
68°	.927	.375	2.475	.404	2.669	1.079
69°	.934	.358	2.605	.384	2.790	1.071
70°	.940	.342	2.747	.364	2.924	1.064
71°	.946	.326	2.904	.344	3.072	1.058
72°	.951	.309	3.078	.325	3.236	1.051
73°	.956	.292	3.271	.306	3.420	1.046
74°	.961	.276	3.487	.287	3.628	1.040
75°	.966	.259	3.732	.268	3.864	1.035
76°	.970	.242	4.011	.249	4.134	1.031
77°	.974	.225	4.331	.231	4.445	1.026
78°	.978	.208	4.705	.213	4.810	1.022
79°	.982	.191	5.145	.194	5.241	1.019
80°	.985	.174	5.671	.176	5.759	1.015
81°	.988	.156	6.314	.158	6.392	1.012
82°	.990	.139	7.115	.141	7.185	1.010
83°	.993	.122	8.144	.123	8.206	1.008
84°	.995	.105	9.514	.105	9.567	1.006
85°	.996	.087	11.430	.087	11.474	1.004
86°	.998	.070	14.301	.070	14.336	1.002
87°	.999	.052	19.081	.052	19.107	1.001
88°	.999	.035	28.636	.035	28.654	1.001
89°	1.000	.017	57.290	.017	57.299	1.000
90°	1.000	.000	Undefined	.000	Undefined	1.000

Index

• Numbers and Symbols •

• A •

• C •

• G •

• H •

• I •

• K •

• L •

• M •

• N •

• O •

• P •

• Q •

• R •

• S •

• T •